Silkworm Physiology

Silkworm Physiology

Mahendra Kamalvanshi

RANDOM

RANDOM PUBLICATIONS

NEW DELHI (INDIA)

Silkworm Physiology

ISBN 978-93-5111-595-3

Published in 2015 in India by

RANDOM PUBLICATIONS

4376-A/4B, Gali Murari Lal, Ansari Road
New Delhi-110 002
Phone : +9111-43580356, 011-23289044, 011-43142548
e-mail: sales@randompublications.com,
info@randompublications.com, randomexports@gmail.com

Reprinted 2026

Type Setting by : Friends Media, Delhi-110089

Preface

The requirement of different nitrogenous compounds in various insects has been investigated. Nitrogen as the major element of amino acids plays an important role in the insect's physiology. The silkworm, Bombyx mori L. derives almost all the nutrients required for its growth from the mulberry leaf. Effects of mulberry leaves supplementation with three nitrogenous compounds on silkworm, Bombyx mori, larvae were analyzed. Dietary supplements of glutamic acid, glycine and urea were fed to silkworm larvae through 1st to 5th instar. The larvae were fed by mulberry leaves of Kokusou variety and the supplementary leaves were used once a day. These compounds showed significant changes in larval parameters such as larval weight and nutritional indices but these nitrogenous compounds could not improve the economic trait of silkworm. In some groups of larvae that the nutritional indices had been improved, the amount of cholesterol and triacylglycerol in the hemolymph increased significantly and the supplementation in these groups have had a great influence on the amount of glucose in the larval hemolymph.

I would like to thank my team for standing beside me throughout my career and writing this book. My special thanks go to "Random Publications" who have published the book.

– Mahendra Kamalvanshi

Preface

Contents

1

Mulberry Silkworm

INTRODUCTION

Heat shock response is a universal physiological phenomenon exhibited by both prokaryotes and eukaryotes. It is normally manifested by the appearance of a set of new proteins or an increase in the quantity of certain specific. Pre-existing proteins in response to exposure of an organism or a cell to thermal- and other forms of stresses.

The heat shock proteins (HSP) are presumed to ensure survival under stressful conditions by involvement in damage protection or damage repair due to their action as molecular chaperons. Generally. the heat shock response depends on the magnitude of temperature elevation and duration of exposure. and is relative to the environmental temperature at which the organism normally survives. Cells which have been Pre-exposed to thermal stress acquire a transient resistance against the killing effect of a subsequent thermal stress and such a thermotolerance is caused by an enhanced resistance of proteins against thermal denaturation and aggregation.

The HSPs are also proposed to protect cells depleted in ATP from rapid necrotic death by inhibiting the aggregation of cytoskeletal proteins. The heat shock response of organisms habitating tropical climate is likely to differ from those from temparate climate. This aspect has direct relevance in such commercially important areas as the field rearing of mulberry silkworms for silk production. The native Indian. multivoltine races of *Bombyx mori* exhibit much more tolerance to high temperatures. often the field temperatures touching 40°C or more in summer. as against the exotic bivoltine races of temperate origin. The latter strains are better yielders of silk fibre in terms of quality and quantity but less popular with the farmers due to their higher susceptibility to environmental fluctuations.

Although temperature and humidity have been known to play important role in silk production. not much information is available on the effect of temperature on silkworm physiology. Most of the heat shock studies have been carried out on cell- and tissue cultures or organisms from temperate climate.

Here we have made a comparative analysis of the effect of heat shock on the tropical and exotic races of silkworm. *B. mori.* at different life stages.

B. mori larvae were reared in the laboratory on mulberry leaves. at 24–26°C and 60–85per cent relative humidity. The silkworm eggs. larvae. pupae and the adults were exposed to temperatures ranging from 37°C to 47°C for specified time. The heat treatment was always coupled to a relative humidity of 90per cent. For every heat shock experiment. at least 20 larvae. 10 pupae or adults. or 100 eggs were used at any time. and each experiment was repeated at least 3 times. After the heat shock. the samples were returned to the rearing temperature and allowed to recover. Survival rate was assessed in the case of eggs by hatchability. larvae by their ability to enter succeeding instar or spin healthy cocoons. pupae by adult emergence. and adult by successful copulation and egg laying.

The effect of exposure to elevated temperature on physiological parameters was analysed. The heart beat rate was counted by examining the dorsal blood vessel of larvae. mounted on plasticin. The protein profiles of the haemolymph. fat body. silk glands and cuticle of the larvae. as well as the total proteins from the pupae and adult moths were analysed by sodium dodecyl sulphate-polyacrylamide gel electrophoresis.

Haemolymph from a cut proleg of the larvae was collected in a precooled graduated centrifuge tube containing a few crystals of phenyl thiourea. The haemolymph was centrifuged at 2000 *g* for 5 min at 4°C. and the supernatant was analysed for protein contents. For the isolation of haemocytes. the haemolymph from 20 larvae was pooled and centrifuged at 1000 *g* for 1 min at 4°C to remove the tissue debris. and the supernatant was centrifuged again at 5000 *g* for 10 min at 4°C. The sedimented haemocytes were examined under the microscope and lysed using the homogenization buffer (200 mM Tris-Cl. pH 7·5; 5 mM MgC12; and 250 mM NaCl). For analysis of proteins. the fat body. silk gland and the cuticle were dissected from the heat shocked- or control larvae. and washed in Ringer.

The tissues were homogenized using a teflon glass homogenizer. NP40 was also included in the buffer while homogenizing the cuticle. The homogenates were centrifuged at 5000 *g* for 10 min at 4°C. and the supernatant after removal of the floating lipid layer was used for protein analysis. Protein content was quantitated by the dye binding method. For *in vivo* labelling. the larvae were heat shocked for 30 min at 41°C and injected with 60 ˜Ci of [35S]Met per larvae. The heat shock was continued for an additional 30 min. The larvae were sacrificed and the tissues processed for protein analysis. The [35S]Met labelled proteins after SDS-PAGE were subjected to fluorography. *In vitro* heat shock studies were carried out by exposing the isolated fat body. cuticle (both maintained in insect Ringer) and haemolymph from control larvae to 41°C for lh.

EFFECT OF HIGH TEMPERATURE EXPOSURE ON THE SURVIVAL OF THE SILKWORMS

The silkworm *B. mori* when exposed to high temperatures ranging from 37°C to 47°C. showed varying levels of heat tolerance at the different developmental stages *viz.*. the egg. larva. pupa and adult. The later larval instars and the pupal stages tolerated higher temperatures for longer durations as compared to the eggs or adults. The data represent the maximum time they could be exposed at any given temperature to give 100per cent survival. The multivoltine *C. Nichi* strain normally reared at 24–26°C. showed complete tolerance for exposures up to 39°C for 2 h. However. on exposure to temperatures of 41°C or above. the survival rates reduced drastically. Beyond 43°C. even a 10–20 min exposure proved lethal to all the developmental stages of the insect. The multivoltine *C. Nichi* and *Pure Mysore* larvae were more resistant than the bivoltine *NB4D2.* when they were exposed to different temperatures for varying times. The differences begin to appear at elevated temperatures of 41–43°C although they were equally tolerant to exposure at 39°C up to 2h. Between the two multivoltine races examined here. *C. Nichi* appeared to be marginally more tolerant than the *Pure Mysore* strain. In the case of bivoltine even when the larvae survived up to the cocoon spinning stage in some instances. they appeared placid. The fifth instar larvae at 3 days showed more thermotolerance than the younger larvae in the 41–43°C range in both multivoltine and bivoltine races when exposed for different times. Therefore. this age of larvae was chosen for all further heat shock studies.

PHYSIOLOGICAL CHANGES

The heart beat of larvae ranged from 54 ± 3 in multivoltine to 58 ± 3 in bivoltine races. On exposure to high temperature. the heart beat rates increased in both races. the most pronounced one being the multivoltine *Pure Mysore* strain. Temperatures of 43°C or more. however. reduced the heart beat rate and the larva become placid. The haemolymph pH (6·4. in all three strains examined) and volume (1·0 ±0·1 ml in *C. Nichi* and *Pure Mysore.* and 1·6 ±0·1 ml in *NB4D2*) did not show variations consequent to heat shock. The total protein contents in different tissues or haemolymph were also not affected by heat shock. The total haemocyte population did not show much variation in the heat shocked larvae. However. the differential count of plasmatocyte and granulocyte showed variations. the former decreasing by 8–12per cent and the latter showing a corresponding increase. in all three strains. The counts of spherulocyte + oenocytoids did not change. The haemocytes from control larvae always showed a characteristic feature of crowding when a droplet of the haemolymph was placed on a glass slide but this character was significantly reduced in haemolymph from heat shocked larvae. The haemocytes of bivoltine *NB4D2* were generally larger than their counterparts from the multivoltine C.

Nichi. The size of granulocyte increased on heat shock in both races. but the enlargement was more prominent in the latter. In *C. Nichi.* on an average the granulocyte diameter increased from 3·5 ± 0·8˜m to 5·1± 0·4˜m whereas in *NB4D2.* the increase was from 5·1 ± 0·6 ˜m to 5·5±0·9˜. In pupae. exposure to high temperature increased the movement of the posterior part of the body. Beyond 43°C. the adult emergence was reduced to 5per cent. The adult moths showed enhanced wing fluttation on increase in environmental temperature and fecundity was reduced to 30per cent by 41°C. The eggs kept above 41°C changed to a brick red colour and the hatchability was reduced to zero.

Protein profiles from heat shocked larvae

The protein profiles of haemolymph from heat shocked larvae in comparison to the control or those allowed to recover from the heat shock. A total of 12–15 protein bands were present in *C. Nichi. Pure Mysore* as well as *NB4D2* strains. The appearance of new 93 kDa protein band consequent to exposure to high temperature was clearly discernible in all of them but their kinetics differed. In *C. Nichi* this protein band appeared immediately on exposure to 41°C and its level remained high during the 2–4 h recovery time at 26°C following the heat shock. In *Pure Mysore* and bivoltine *NB4D2* strains. the 93 kDa protein band could be seen only after 2 h recovery from thermal shock. There were no noticeable changes in any of the other protein bands. Exposure of larvae to temperatures up to 39°C for 1 h did not result in any change in the haemolymph protein profiles compared to the controls. The appearance of the 93 kDa protein band in the haemolymph of *C. Nichi* was confirmed also by *in vivo* [35S]Met labelling studies. In contrast to the *in vivo* result. when the haemolymph isolated from control larvae was exposed to 41°C for 1 h the heat shock resulted in protein degradation. Since the HSPs are known to be nonsecretory. the appearance of a new protein band in haemolymph was unexpected. In order to see whether this protein band had arisen from the haemocytes which populate the haemolymph. the protein pattern of the isolated haemocytes from such heat shocked larvae was examined. Only a few protein bands were seen in haemocytes but the appearance of a new 70 kDa protein band was conspicous in *C. Nichi* haemocytes. 2 h post recovery from thermal shock. This protein band was clearly absent in *NB4D2* haemocytes. However. the 93 kDa protein band was not seen in either race.

HEAT SHOCK PROTEINS FROM FAT BODIES

The protein profiles of fat bodies derived from control larvae and those exposed to 41°C for 1 h. of multivoltine *C. Nichi* and bivoltine *NB4D2* races. As seen in the haemolymph. the most characteristic change was in the appearance of the 93 kDa band in both *C. Nichi* and *NB4D2* following heat shock. In addition. a 89 kDa band was also observed. In *C. Nichi.* the 70 kDa protein was detectable in the control larvae. which increased on heat shock and stayed at high levels during the entire recovery time. Although the 70 kDa band was apparent also in the bivoltine *NB4D2.* the changes in its level on heat treatment did not vary as conspicously as in *C. Nichi.* The fat body proteins from larvae heat shocked at 39°C also did not show any change in their profiles. Subjecting the isolated fat bodies from control larvae to heat shock for 1 h at 41°C. again resulted in protein degradation as was observed in the case of haemolymph. *In vivo* labelling with [35S]Met during heat shock confirmed the 89 and 93 kDa proteins in fat bodies of *C. Nichi* although no prominent change was apparent in the level of 70 kDa protein.

CUTICULES

The protein profiles were more complex due to the presence of larger number of protein species. Nevertheless. in C. *Nichi.* the presence of additional protein bands at 93 kDa. 46 kDa and 28kDa immediately on exposure to 41°C was distinct. The 70 kDa protein band was clearly seen but did not show much change in the level on heat shock. The appearance of these protein bands in *C. Nichi* following heat shock was also confirmed from the *in vivo* [35S]Met labelling studies. These changes were similar in cuticle proteins from the *Pure Mysore* and *NB4D2* strains.

SILK GLAND PROTEINS

The silk gland proteins from the heat shocked larvae did not show any clear changes in the protein profiles. Fifth instar larvae start synthesizing large quantities of proteins from third day onwards and these proteins could have masked the HSPs. Alternatively. the absence of HSPs could be attributed to the silk glands not sensing the shock due to their location within the larvae. covered by haemolymph. fat body and cuticle. Due to such insulation. the heat stress may not even reach the silk glands. when the whole larvae is subjected to heat shock. The absence of HSPs in the silk gland could be a consequence of this and may not reflect the inability of the tissue to synthesize them. However. it should be noted that the haemocytes from *C. Nichi* though not *NB4D2* showed the presence of HSP.

HEAT SHOCK RESPONSE IN PUPAE AND ADULT MOTHS

The protein profiles of pupae and the adult moths of *C. Nichi* when exposed to 41°C for 1 h. At both stages. the 93 kDa proteins appeared immediately on exposure to the high temperature. In adult moth. additionally the appearance of a 56 kDa band was clear. Further. some of the low molecular weight protein bands nearly disappeared after heat shock of the moth.

DEBATE

Exposure of an organism or tissue to temperatures higher than the optimal growth temperature disturbs their normal function. ultimately resulting in death. An early consequence of heat shocking is the appearance of a set of highly conserved proteins. the HSPs. ubiquitous to all organisms examined so far. The HSPs are considered to confer protection against the adverse effect of heat shock. The susceptibility of economically important organisms such as the silkworms to environmental fluctuations of temperature. humidity etc.. attain significance in their field-rearing for the commercial production of silk. The three strains of *B. mori* examined here. *viz.*. *C. Nichi* and *Pure Mysore* of the multivoltine race. and *NB4D2* of the bivoltine race show different levels of tolerance for exposure to temperatures higher than their normal growth temperature. *C. Nichi* was the most tolerant of the three while the exotic strain *NB4D2* was most labile. as judged by their survival and capacity to spin cocoons after a 1 h heat shock. Temperatures above 43°C proved lethal for all of them. These observations are in general agreement with the commercial field rearing experience on the silkworms. The sensitivity to thermal exposure was also dependent on the life stages of the silkworm.

The fifth instar larvae and the pupae were more tolerant compared to the young larvae or the adult moths. The eggs were also very sensitive to temperature of 39°C and above. when the hatchability was reduced to zero. Having assessed the differences in thermal tolerance of the various races. the effect of heat shocking the whole larvae on various physiological parameters. and the protein profiles of individual tissues or haemolymph of *B. mori* were analysed on a comparative basis. The approach used here is different from those attempted previously which were confined mostly to cells maintained in cultures. In haemolymph. fat bodies and cuticle. the appearance of a 93 kDa protein consequent to heat shock was evident in the multivoltine and bivoltine races of silkworms analysed. However. there were differences in the time of their appearance. whether immediately on heat shock or after permitting them to go through a recovery period.

The protein band made the appearance instantaneously after the heat shock in haemolymph of the more thermotolerant *C. Nichi* compared to the bivoltine *NB4D2* where the protein appeared only after 2 h. during recovery. The 70 kDa protein was present constitutively in fat body and cuticle of both races. On heat shock. the levels of 70 kDa protein showed substantial increase in the fat bodies. from *C. Nichi.* The fat body is the reservoir and active synthetic tissue for most of the haemolymph and cuticular proteins. Only a small number of haemolymph proteins are synthesized by the haemocytes. In the haemocytes from heat shocked *C. Nichi.* only the 70 kDa. was heat induced but the 93 kDa protein was absent. There were no signs of haemocyte lysis or damage consequent to heat shocking. Since the 93 kDa protein is seen after heat shock in the haemolymph. fat bodies and the cuticle. it is likely that the protein is synthesized in the latter tissues and found its way into the haemolymph. But so far. none of the HSPs have been shown to be secretory in nature. The haemolymph being an open circulation system most of the tissues including the fat bodies and silk glands lie bathed in haemolymph within the *B. mori* larvae. Consequently several proteins synthesized in the fat body find their way into the haemolymph. Even Non-secretory proteins from the fat bodies may thus enter the haemolymph due to the immersion of the former in the haemolymph. This is a most likely reason for the presence of 93 kDa protein in haemolymph. The differences in the pattern of expression of HSPs in different tissues of silkworm. seen here are quite unusual.

A complex regulatory pattern is seen in the case of *hsp* gene expression in the flesh fly. *Sarcophaga crassipalpis.* Here the expression of the different HSPs in different tissues vary depending on the stage of development or even the temperature at which the exposure was given. Coulon-Bublex and Mathelin had reported the appearance of a 70 kDa stress response protein after heat shock or acid treatment of the diapausing embryos of *B. mori.* which was also present during deep diapause. HSPs provide protection to cells against the heat induced damage and the synthesis of HSPs is related to heat resistance. Thermotolerance is expressed as an attenuation of damage formation or as a better repair of protein damage. termed facilitated recovery. The total concentration of HSPs (both constitutive and inducible) and their redistribution to specific subcellular sites are considered as more important factors in the acquisition of thermotolerance.

In this respect. the presence of HSP in the haemolymph of *B. mori* can be considered as a desirable feature in conferring thermotolerance to the larvae. Integuments consisting of the cuticle and the underlying epidermis are the external tissues which act as initial Defence organ in most insects. Response of the integument to heat shock will therefore be an integral part of the survival of the silkworm. The appearance of 93 kDa. 70 kDa. 46 kDa and 28 kDa proteins was conspicous in all strains examined consequent to exposure to high temperature. The small molecular weight HSPs in cuticle of *Manduca sexta* have been speculated to provide an alteration in the relative water loss allowing evaporative cooling during heat shock. Our results here show that *in vitro* heat shocking of the isolated tissue or haemolymph at 41°C from control animals reared at 24–26°C. generally resulted in protein degradation whereas subjecting the whole larvae to thermal shock for an identical time and temperature. resulted in the appearance of the additional protein bands. The synthesis of these proteins *in vivo.* in response to heat shock may therefore play important role in the thermal tolerance of the silkworms.

IMPACT OF FEEDING SELECTED MULBERRY GERMPLASM VARIETIES ON SILKWORM

About 92.20per cent of the silk produced in the world is obtained from mulberry silkworm *Bombyx mori* L. reared solely on mulberry leaves. It is well-established fact that in sericulture. more than 60per cent of the total cost of cocoon production goes towards mulberry production alone. Hence. in recent years maximum attention has been given for the improvement of mulberry in terms of both quality and quantity. Silkworm is essentially monophagous insect feeds solely on mulberry leaves.

Leaf quality is an important parameter used for evaluation of varieties aimed at selection of superior varieties for rearing performance. Growth and development of silkworm *Bombyx mori* L. is known to vary depending on the quality and quantity of mulberry leaf used as food source. which in turn indicated by commercial characteristics of cocoon crop. Superiority of different mulberry varieties used as food for silkworm larvae greatly affects the economy of sericulture industry. Nutritive value of mulberry leaf is a key factor besides environment and technology adoption for better growth and development of the silkworms and cocoon production. It is a confirmed fact that. leaf quality differs among mulberry varieties which in turn responsible for the difference in silkworm rearing performances. Leaves of superior quality enhance the chances of good cocoon crop. In the present study an attempt has been made to evaluate better performing mulberry variety through silkworm rearing for Kolar district. one of the premier and traditional sericulture belt in Karnataka. accounts for 40per cent of total raw silk production.

Mulberry plants grown in garden and silkworm rearing experiments were conducted in rearing house at Bethamangala. Bangarpet taluk in Kolar district. Karnataka. In the present experiment mulberry varieties *viz.*. S1708. MS5. C6. C10 and M5were used. M5 mulberry variety is used as a check variety. Selected mulberry varieties cuttings were procured from Central Sericultural Germplasm Resources Centre. Hosur. Tamil Nadu. India and disease free cross breed silkworm egg layings obtained from National Silkworm Seed Project. Bangalore. Karnataka were used. Experiment was conducted in RBD method with 4 replications/ variety. Two years old plants were used for silkworm rearing from time to time in different seasons *viz.*

SILKWORM MOULTING

Moulting test conducted up to 2nd moult following standard rearing methods suggested by Krishnaswami with 4replications/variety and 100larvae/ replication. Healthy tender leaves harvested and fed to young age silkworms up to 2nd moult. Daily three feedings were given at 7am. 2pm and 10pm from brushing to end of II moult. First appearance of one larva out of moult was considered as commencement of moulting. Moulting ratio with respect of all the varieties under evaluation was fixed depending on time duration which recorded more than 50per cent of the larvae under moult. Larval weight was also recorded.

SILKWORM REARING

Silkworm rearing experiments were conducted at different seasons. For each mulberry variety. one egg laying was reared and 4 replications were maintained. After III moult. 100 larvae/replication were maintained. Appropriate cellular rearing techniques were adopted and separate rearing trials were conducted for different varieties. Larvae were fed three times daily with healthy. fresh leaves. Young age larvae were fed with tender. succulent and nutritious leaves known to favour growth and development of chawki silkworms. while mature and coarse leaves were fed to late age silkworms till ripening. Cocoons were collected on 5th day of mounting and assessed for commercial parameters *viz.*. ERR. cocoon weight. shell weight. shell percentage. filament length. reelability. denier and renditta by Sonwalkar methods.

STATISTICAL ANALYSIS

Data collected on various parameters were tabulated and subjected to critical statistical analysis by adopting 'Method of Analysis of Variance' appropriate to the experiment.

CONSEQUENCES AND DELIBERATIONS

Studies on the influence of different mulberry varieties on silkworms' Behaviour and cocoon traits were studied in tropical conditions of India. It was also emphasized that. mulberry leaf quality has direct effect on food consumption ratio. larval growth. digestive coefficient and food absorption. Significant differences were observed in larval weight and moulting ratio. Silkworms reared on S1708 leaves revealed highest larval weight and moulting ratio and lowest larval weight and moulting ratio were observed in C6 leaves during I moult. In II moult. high larval weight and moulting ratio was observed in silkworms reared on S1708 leaves and lowest larval weight and moulting ratio was recorded in C6 leaves. It is evident that tender. succulent and nutritious leaves are known to favour good growth and development of young age silkworms whereas progressively mature leaves with less moisture content are required for late age silkworms.

Silkworm Moulting

Degree and uniformity of moulting varies with quality of mulberry leaves fed. Superior quality leaves favours higher moulting. better growth rate and weight of silkworms. Bongale and Chaluvachari opined that. lower larval weight and moulting ratio in Mysore local variety were associated with lower leaf moisture content and moisture retention ability. Chaluvachari and Bongale reported that S41 variety with higher protein and lower sugar content encouraged high larval weight and low moulting ratio. Mishra *et al*.. recorded 89.16per cent and 92.82per cent moulting ratio in PMxNB4D2 and NB18xNB7 races respectively with S54 mulberry genotype. Sujathamma *et al*.. reported that. when CB and BV larvae fed on Tr10 and MR2 mulberry varieties shown higher moulting ratio and larval weight. Mallikarjunappa *et al*.. observed the superiority of S30. S36 and Vishwa mulberry genotypes over M5 genotype in moulting ratio and larval weight.

Yogananda Murthy *et al*. reported that. mulberry varieties S1708. Tr8 and Tr12 were superior among the ten varieties examined for bivoltine silkworms in moulting tests. Veerapura Narayanappa Yogananda Murthy*et al*.. observed that. among six mulberry germplasm varieties *viz*.. Tr8. Tr12. Tr20. *Matigara black*. *Morus nigra* and M5. in moulting test. crossbreed silkworms reared on Tr8 mulberry leaves recorded highest values and lowest values recorded in silkworms reared on *Morus nigra* leaves both in I and II moult respectively.

Silkworm Rearing

Significant differences were observed in larval parameters and commercial cocoon characters. Ten 5th instar larval weight was significantly higher in silkworms reared on S1708 leaves and lower larval weight was recorded in silkworms reared on C6 leaves. Single cocoon weight was higher in cocoons of silkworms reared on S1708 leaves. However. significantly lower cocoon weight was registered in silkworms reared on C6 leaves. Both shell weight and shell percentage were significantly high in cocoons of silkworms reared on S1708 leaves while cocoons obtained from silkworms reared on C6 leaves recorded lower shell weight and shell weight percentage respectively. Filament length and reelability percentage was significantly high in cocoons obtained from silkworms reared on S1708 leaves. Lower filament length and lower reelability was recorded in cocoons harvested from silkworms reared on C6 leaves.

Significantly lower renditta was recorded in the cocoons procured from silkworms reared on S1708 leaves. while higher renditta was found in the cocoons recovered from silkworms reared on C6 leaves. Finer denier was recorded in cocoons procured from silkworms reared on S1708 leaves while cocoons procured from silkworms reared on C6 leaves produced coarser denier. Effective rate of rearing was significantly higher in silkworms reared on S1708 leaves and lower effective rate of rearing was recorded in silkworms reared on C6 leaves.

Krishnaswami *et al*.. observed that Berhampore variety was better than Kosen and Mandalaya with regard to effective rate of rearing and cocoon weight. Verma and Kushwaha reported that. mulberry variety Catteneo was found best in silkworm rearing trials compared to Burmose2. Tsukasakhu and Local mulberry varieties. Govindan *et al*.. observed that. cocoon weight obtained with S41 and S54 was higher than that obtained with S36 while Mysore local and Kanva2 registered lowest. Venugopala Pillai *et al*.. reported that mulberry variety S54 encouraged higher values in larval span. larval weight. single cocoon weight and coon yield. Tayade *et al*.. reported that mulberry variety S54 was found superior followed S41 and Kanva2 for feeding silkworms.

Dar *et al*.. observed that feeding of Ichinose leaves resulted in higher cocoon characters compared to other two varieties. Fotadar *et al*.. observed that silkworms fed with Kokuso27 variety leaves revealed best results among other varieties studied. Sathyanarayana Raju *et al*.. stated that mulberry variety S30 showed better performance than S36. S41 and K2 for commercial characters of bivoltine cocoons. Saratchandra *et al*.. revealed that. mulberry variety S36 was found superior in silkworm rearing trials and recorded highest cocoon yield. ERR and shell percentage. Changalarayappa and Chinnaswamy. Vage and Ashoka reported that. silkworm breeds and performed well when reared on M5 mulberry variety. Rahman *et al*... Chakrovorty and Borgohain. Santoshagowda V. Patil were reported that. mulberry variety S1635 was found superior in silkworm rearing trials with good commercial cocoon characters compared to all other varieties examined. Rachotaiah *et al*.. noticed that. mulberry variety RFS175 was found superior in silkworm rearing trials with maximum cocoon production. Sujathamma *et al*.. observed that. mulberry varieties Tr10 and MR2 were found superior as they secured maximum scores for silkworm races CB and BV in rearing tests.

Definite relationship between larval weight and economic characteristics such as cocoon weight and shell weight in various strains were reported. Adolkar *et al*... Giridhar and Reddy. Seidavi reported that. silkworm rearing performance differed significantly when subjected to same conditions. some of them performed better and poor performance by some races. Present study confirms the same as S1708 mulberry variety gives better results in pre-cocoon and post-cocoon characters compared to other varieties tested. Gangawar reported that. among eight mulberry varieties *i.e.* S1. S146. S1635. AR12. AR14. TR10. BR2 and K2 evaluated for nutritional potential by rearing experiments. silkworms fed on BR2 variety showed high larval weight. cocoon weight. shell weight and silk percentage in comparison with other varieties. Yogananda Murthy *et al*.. observed that. mulberry varieties S1708. Tr8 and Tr12 were superior among ten varieties examined with bivoltine silkworms for growth and development of silkworms and commercial cocoon parameters. Veerapura Narayanappa Yogananda Murthy*et al*.. opined that. among six mulberry germplasm varieties *viz*.. Tr8. Tr12. Tr20. *Matigara black*. *Morus nigra* and M5 observed for leaf quality. crossbreed silkworms reared on Tr8 mulberry leaves proved significantly better and lower values observed in silkworms reared on *Morus nigra* leaves. Ogunleye and Johnson evaluated three silkworm races namely EC1. EJ1 and EJ2 for their growth and productivity by feeding them with S36 mulberry leaves and reported that. EJ2 silkworms showed a higher and consistent growth rates compared to other silkworm races. Cocoon weight and shell weight are the most important characters evaluated for productivity. Shell percentage indicates the amount of raw silk reeled from given quantity of fresh cocoons and shell percentage varies according to silkworm age and breed. According to FAO. total silk filament length is ranging from 600m-1500m out of which only 80per cent is reelable. In the present study. silk filament length of cocoons falls within this range and cocoons recovered from silkworms reared on S1708 mulberry leaves produced longest filaments length and lowest denier.

FINALES

In 1980s. mulberry varieties like S36. S41. S46. S54 etc.. and in late 1990s varieties like S1. S146. S1635. AR12. AR14. TR10. BR2 etc.. were evaluated for nutritional potential and silkworm rearing. In recent years more new mulberry varieties like S1708. MS5. C6. C10. were evolved and screened to know their nutritional quality of leaves. Mulberry being the sole food plant of silkworm *Bombyx mori* L. and nutritional quality of leaves evaluated through moulting and bioassay tests is a standard recommended practice. Present results revealed that. moulting and rearing performance of silkworms proved better with S1708 mulberry variety and cocoons harvested showed significantly good economic cocoon characters such as silk filament with finer denier and lower renditta compared to other varieties. Results thus obtained are presented in this paper which is self-explanatory. Information generated will help silkworm breeders and rearers at field level to use these new mulberry varieties for high yield. healthy silk worms growth and development and good commercial cocoons parameters for sustainable growth and development of sericulture industry.

BIOLOGICAL AND TECHNOLOGICAL EFFECTS OF MULBERRY

The quality of mulberry leaves plays an important role in the success of the sericulture industry and directs its economics. and hence much effort and research have been carried out to improve the quality and quantity of mulberry-leaf production for silkworm rearing and then cocoon production. Some varieties of mulberry leaves appeared to be superior to others. It is well-known that the amount and quality of mulberry leaves affects growth rate. developmental period. body weight. and survival rate of larvae. as well as influencing the subsequent fecundity. longevity. movement and competitive ability of the adults. The nutritional status of mulberry leaves can be improved by enriching them with extra nutrients to increase larval growth and improve cocoon characteristics.

Mulberry leaves immersed in solutions of 0.5. 1.0 and 1.5per cent ascorbic acid solution significantly increased the weights of both larvae and cocoons of silkworm compared with untreated leaves. Suprakash and Pal supplemented fresh mulberry leaves with 3 levels of vitamin B complex by dipping them in 0.5. 1.0 and 1.5per cent solution. and fed the dried leaves to larvae of various

Bombyx mori races: the 0.5per cent level increased the weights of larvae and cocoons. and the shell ratio. Treatments of *B. mori* larvae with propolis extracts yielded heavier cocoons and cocoon shells. a higher silk content. and females laid more eggs than controls. Zannoon reported that. Various solutions of bee honey increased the mature weight of the larvae. silk gland. fresh cocoon and cocoon cortex. the number of deposited eggs per female. gave heavier. longer filaments of reeled cocoons. but did not affect cocoon silk content ratio and filament size. We evaluate here the efficiency of two varieties of mulberry leaves compared with *Morus alba* var. Balady. and determine the best feeding schedule. We then study the effects of fortification of mulberry leaves by vitamin C. vitamin B and three kinds of bee-honey on the biology and silk production of mulberry silkworm *Bombyx mori*. Widely cultivated in Japan.

Morus alba var. Kokuso-27 was generated by an artificial hybridization between Naganuma. Its leaves are characterized by their shiny dark green colour. by remaining soft for a long time. and by tolerating drought. It can also be grown under wet conditions. It has high water. and total sugar 14per cent content. *Morus indica* Kanva-2 is diploid. and widely cultivated in southern India. a result of selection from a natural population from Mysore. It has lower leaf moisture and sugar content. but a high rooting ability and great adaptability. It is resistant to leaf spot. and moderately resistant to leaf rust and powdery mildew. The rearing of silkworm (*Bombyx mori* 9F 7X- Chinese hybrid) was carried out under laboratory conditions (26±2 °C. 70±5per cent relative humidity). according to the technique of Krishnaswami. The experiment had two parts. Firstly. larvae were reared on fresh mulberry leaves of three varieties (*Morus alba* var. Kokuso-27. *Morus indica* var. Kanva-2 and the native *Morus alba* var. Balady) under three different feeding schedules (two. three or four feeds per day).

Second. larvae were reared on the variety of mulberry identified as the best one from the first part. enriched with 1per cent vitamin C. 0.2per cent vitamin B or one of three kinds of bee-honey (clover. cotton and citrus. each at 50per cent: each was compared with the control group. Mulberry leaves were dipped in each material for one minute and left to dry. and fed to larvae once per day; control leaves were dipped in distilled water. Each group was replicated three times. The following biological variables were recorded. During development. larvae need large amounts of mulberry leaves costing effort and money. and hence the duration of the larval and moulting period (D) was measured for each tray. During each of the fourth and fifth larval instars. the daily number of dead larvae was recorded from each tray. and larval mortality rate (m) calculated as a percentage of the total number of larvae alive at the beginning of the 4th instar. The cocooning percentage was calculated as the number of fresh cocoons produced as a percentage of the total number of larvae alive at the beginning of 5th instar. Six pairs of adult moths from each tray were paired separately in paper cages. and the total number of deposited eggs per female were counted as a measure of fecundity (V). The hatchability percentage was the total number of hatched eggs per female as a percentage of the total number of eggs laid. Individual fitness (r) was calculated as a integrated performance measure by combining development time and fecundity (V) via the equation:

- r = [Ln]/ D
- where Ln = natural logarithms. m = survival probability. V = fecundity. D = duration.

A number of technological variables were also recorded. For coccon weight. fifty cocoons from each group were collected randomly soon after pupation. cleaned and weighed separately. The silk content ratio was measured from the weight of the cocoon shell (weighed after removing the internal pupa) as a percentage of the weight of the fresh cocoon. For the size of a reelable filament (dn). fifty cocoons from each group were collected. oven dried and reeled by an individual reeling machine. and the reelable filament length and weight recorded; the size of reeled silk filaments was calculated as the weight of the silk filament divided by the length and multiplied by 9000. Biological and technological data were statistically analyzed using Anova implemented by the Coastat software Programme. Of the three mulberry varieties tested. larvae fed on Kokuso-27 with feeding schedule four feeds per day showed the shortest total larval duration the lowest total mortality rate. and the highest larval fitness. Similarly. when fed Kokuso-27 with four feeds per day. a significant increase in cocooning percentage. fecundity and egg hatchability resulted. In virtually all cases the Balady variety was intermediate and the Kanva-2 variety poorest. The analysis of interaction between effects of mulberry varieties and feeds number/day revealed Non- significant variation within biological characters except fecundity and hatchability. There were significant increases in fresh cocoon weight. silk ratio and filament size in larvae reared on Kokuso- 27. with those fed four feeds per day having the highest values. As before. larvae fed on Kanva-2 recorded the poorest values for cocoon and filament characters. with Balady usually in between. There were significant variation within means of technological characters as response to mulberry varieties. feeds number/day and the interaction between them ($F4.441 = 3.31. p<0.05$ and $F4.441 = 3.69. p<0.01$). except the analysis of interaction of filament size revealed non significant variation. Feeding larvae with mulberry leaves enriched with additive compounds improved the biological measures in all tested groups compared with the control.

There was a significant lowering in larval mortality rate compared to the control. especially in clover and cotton honey groups. Vitamin C. clover and citrus honey groups showed a significant enhancement in cocooning percentage. All test groups recorded higher fecundities and hatchability percentages than the control. but there were no significant differences in larval duration and fitness. In technological measures. there were a significant increases in cocoon weight and silk ratio in all tested groups. especially vitamins C. B and clover honey compared with the control. The reeled silk filaments of all tested groups exhibited higher filament size than the control group. Rearing *Bombyx mori* on *Morus alba* var Kokuso-27 leaves with four feeds per day clearly improved the rearing and silk productivity of the tested larvae compared with the control. whilst rearing larvae on the Kanva-2 variety had a deleterious effect.

The superiority of Kokuso-27 might be explained by its increased amounts of crude protein. soluble sugar and starch content in their leaves. which might improve the health and growth of larvae. perhaps by increasing the rates of the digestive and oxidizing enzymes which help in utilizing food and increasing food consumption. Zannoon reported that rearing grown larvae with Kokuso-27 led to shorter larval durations and lower larval mortality rate relative to feeding on the native Balady variety. The improvement in cocooning percentage. fecundity and hatchability in the Kokuso-27 treatment might be referrable to haemolymph protein improvement and an increase in storage proteins. This improvement might also be due to levels of bombyxin in the haemolymph. This hormone is modulated by the brain in response to variation in nutrition. and forms part of the mechanism coordinating the growth of internal organs with overall somatic growth; it also plays a role in carbohydrate metabolism. In technological measures. the Kokuso-27 treatment increased fresh cocoon weight. silk content ratio and silk filament size. especially with four feeds per day.

The Kanva-2 again exhibited the poorest results. The increase may be due primarily to the promotion of silk protein synthesis by increased enzymatic activity of aminotransferase. involved in the uptake of nitrogen from leaves by body tissues and the silk gland: Machii and Katagiri found a high correlation between leaf protein levels and the production efficiency of the cocoon shell (*i.e.* cocoon shell weight relative to total consumption). Dai *et al.* indicated that ecdysone plays a significant role in nucleic acid metabolism and the related protein synthesis in silkworm: it induces both growth and silk production. Ashour tested differences among three different mulberry varieties (Kokuso-27. Morittiana and Rosa). and also showed that Kokuso-27 had the most efficacious effect on silkworm production. Enriching mulberry leaves with additives led to an obvious improvement in several measures. lowering larval mortality rate and elevating cocooning percentage. fecundity and hatchability. but not affecting larval duration or fitness. The enhancement of larval development may be due to the action of vitamin C and B as coenzymes in amino-acid metabolism and antioxidant agents. which may increase amino-acid concentrations in larval tissues. leading to improvements in productivity. This hypothesis coincides with that of Babu *et al.*. who observed that leaf consumption was greater in ascorbic-acid-fortified leaves as compared to control: tissues and body fluids contained amounts of vitamin C dependent upon the amounts ingested in food. Suprakash and Pal also found that vitamin B complex significantly improved growth and development. with beneficial effects on the economic characteristics of the cocoon.

The improvement of cocoon and silk characters in this experiment may be attributed to improvement of the efficiency of conversion of dietary nitrogen into the cocoon shell. In other words. the increase in the cocoon and filament characters might be due to increased protein conversion efficiency of the silk glands as a result of increased availability of vitamins and the positive antibacterial effect of honey on the growth of many kinds of bacteria. This is supported by Sarker *et al.*. who reported that supplementation of mulberry leaves offered to silkworm with ascorbic acid (1per cent) and vitamin B complex improved cocoon yield and silk filament quality. Basson *et al.* assumed that the antibacterial activity of honey was attributed to multiple factors such as pollen present in honey which could be the source of the antibacterial aromatic acids. The hypertonic sugar concentration may also play an important role in this activity. Nafea reported that clover and cotton honey show antibacterial activity: clover and citrus honey were better than cotton honey. probably because of the lower levels of fructose and glucose in cotton honey. According to the overall results of this study. the mulberry variety Kokuso-27 is highly recommended for rearing silkworm in Egypt. and a feeding rate of four feeds per day is also recommended. because of its beneficial effects on silkworm biological and economic productivity. Fortification with either vitamins or bee-honey is also highly beneficial. improving rearing and the quality of the silk filament.

2

Non-Mulberry Silk Worms

INTRODUCTION

The natural silks are broadly classified as mulberry and wild or Non-mulberry. Non-mulberry sericulture is universally known as forest or wild sericulture. Tropical and temperature tasar. eri. muga and anaphe are the principle non-mulberry silks. Other varieties *i.e.* fagara. coan.mussel and spider silks are limited interest. Nearly 95per cent of the global production of non-mulberry silks is tasar. All branches of sericulture require food plants and manpower. In mulberry sericulture over 60per cent of the cost of production goes into raising and maintaining the food plants. besides a heavy initial investment is necessary for rearing houses. rearing appliances and other essentials. Like wise. among the non-mulberry varieties. eri has the disadvantage of higher production costs.

Tasar is endowed by nature with vast potential. But continuous deforestation resulted in depletion of non-mulberry food plants. One should not forget that non-mulberry sericulture holds great promise for the world forestry as a supplementary activity. Non-mulberry sericulture is a forest-based industry uniquely suited to the economy and social structure of developing countries because of its minimum investment requirement. high employment. and foreign exchange earning potential. In India about 12.43 million hectares of unexploited tasar food plants in the tropical and temperate belts could be put to use without investment. Moreover. unlike large industries in which heavy investment is indispensable and the employment potential is limited. tasar culture requires a very low investment for net high returns. Various operations of tasar production can be carried out even in remote forest villages as they do not require electricity. complex machinery. specialized skills etc. Non-mulberry production offers rural population an attractive source of income. arrest their migration to urban areas. thereby preserving their traditional skill and way of life.

DISTRIBUTION

A large number of species: are used in the production of Non-mulberry silks. But only eight have been commercially exploited in Asia and Africa tribal communities. They are Tasar silk. Muga silk. Eri silk. Anaphe silk. Fagara Silk. Coan silk. Mussel silk and Spider silk. Non-mulberry silks are commercially produced mainly in China and India. However major share of production (80per cent) goes to Chine. Climate and vegetation provide India with an abundance of non-mulberry sericigenous fauna. India produces significant amounts of all the important varieties of non-mulberry silks. tropical and temperature tasar. eri and muga.

TASAR SILK

These are reared in the tropical and temperate zones. Four species fo the genus *Antherea:* are used for commercial production. They are as follows.

- Tropical tasar silkworm
  - *A. Mylitta*. D
- Temperate Silkworm
  - *A. Proyli.* J.
  - *A. Pernyi.* G.M.
  - *A. Yamamai*.

Antherea comprises more species than any other genus of sericigenous insects. So far thirty five species have been recorded (31 in the Indo-Asutralian biogeographic region. 3 in Palearctic region. one in USA). These are uni or bivoltien types.

Tropical Tasar:

These are in dense. humid tropical forest fo central and southern platead of India. The major cocoon-producing districts are Sginghbhum and Santhal Pargana in Bihar. Raigarh and Jagdalpur in Madya Pradesh; Mayurbhanji and Keonjhar in Orissa; Purulia and Bankura in West Bengal; Bhandra in Maharashtra; Adilabad. Warangal. Karimnagar. Khammam. Mahaboobnagar. Visakapatnam of Andhra Pradesh; Belgaum in Karnataka. A. mylitta feeds on Teminalia and produces a special type of silk.

The life cycle and rearing is as follows.:Eggs are oval. dorsoventrally flattened and bilaterally symmentraical along the enteroposterior axis. Each measures 3mm in length and 2.5mm in diameter. weighing 10mg. eggs are white. light yellow or creamy. Two brownish parallel line along the equatorial plane of the egg devide the surface into three zones:. Eggs hatch in 3-5 days. Larvae are eruciform and possess a hypognatus head with biting and chewing mouth parts.

Newly hatched larva is dull brownish yellow with black head. measures 7 x 1 mm and weighs about 8 mg. after 48 hrs. larval body turns green and head becomes brown. Occationally yellow. blue and almond coloured larvae are also found. Mature larva weighs 50 grams and measures 13 x 2.1 cm. each larvae passes five instars. The larval duration is 26-28 days:; 42- 45 days:; 55-60 days:. The prothoracic hood of the first instar larvae dorsally bears an oval black spot. which early in the seconds instar becomes M-shaped. later on V-shaped with two dots. These are absent in third instar. but appear in the fourth and fifth instars as two semilunar red markings.

The anal flap has a triangular black mark early in the first instar. which becomes V-shaped and brownish from second instar onwards. Sexual marking appear lat in the fifth instars as milky white spots on the ventral surface of the eighth and ninth abdominal segments in all instars the larvae possess five types of tubercles (dorsal. upper lateral. lateral. lower lateral. caudal).

They are black in the first instar. orange red in the second and violet in the third to the fifth. White. minute hair are distributed irregularly all over the body. Setae are of two kinds. Silver shining spots appear during the third instar of second to seventh abdominal segments. Pupa is obtect adectious with segmented. dark brown colour body measuring 4.5x2.3 cm and weighs 10.3 grams. The sexual markings are on eighth and ninth segments. Cocoon is single shelled. pendent. oval. closed and reelable. non-flossy with fine grains the anterior end has dark brown peduncle with a ring at the distal end. Cocoons are yellow or grey. Moths exhibit sexual dimorphism.

The females are bigger: with a bipectinate antennae: and broad abdomen. Males are smaller: with broad antennae and narrow abdomen. Male wing span is about 16 cm. the area of fore wing is about 2121 mm2 with a centrally positioned ocellus:.

While hind wing is about 1584 mm2 with an ocellus of 50mm2. Female wing span is 18 cm. The fore wing is about 2350 mm2 with ocellus:. The hind wing is 1850mm2 size with ocellus: The colour of ocelli is same in both sexes. Wing scales. wing venation are specific. The last four abdominal segments: are modified to form the genetalis.

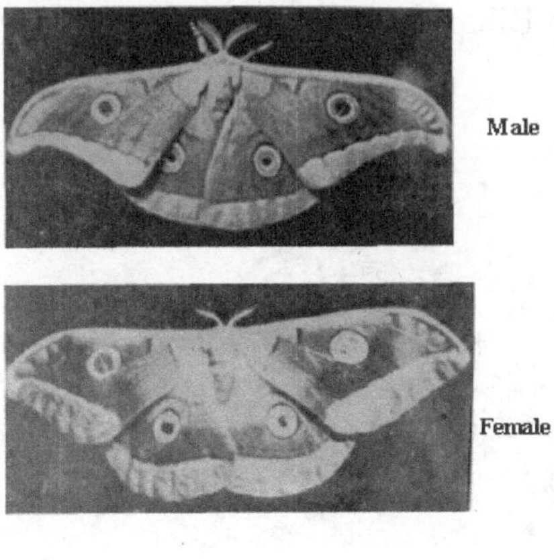

Fig. Amylin Moth

Tasar silkworm is polyphagous. The primary food plants are eight types. Besides more than two dozens of secondary food plants.

1. *Terminalia tomentosa*. W and A.
2. *T. arjuna.* WandA.
3. *Shorea robusta*. Roxb
4. *Lagerstroemia parviflora*. Roxb.
5. *L. Speciosa*. Pers.
6. *L. indica*. Linn.
7. *Zizyphus mauritiana*. Lam.
8. *Hardwickia binata* Roxb

T. tomentosa *T. arjuna*

Fig. Food Plant

Outdoor rearing causes 50-5per cent: loss besides 35- 40per cent loss due to diseases. pests. Hatched larvae are placed on the Terminalia leaves. Since the larva are exposed to natural conditions. the fate of the crop largely depends on choice of rearing site and food plants. brushing. supervision and maintenance of larval population and other rearing operations. The cocoons are collected from the branches of food plants and marketed. The produce is measured in "kahans": and sold in local numerical lots. which are as follows.

- Ganda
- Pan
- Kahan (1280 in Bihar. 1600 in Orissa. 1000 in MP)
- Khandi

Fig. A. Outdoor Rearing B. Grainage

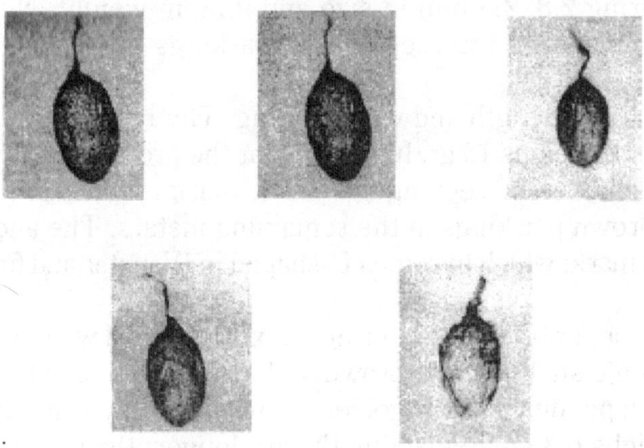

Fig. Cocoons of Temprate Tasar

This is an hybrid worm. The oak tract extending from Jammu and Kashmir in the West to Manipur in the east. embracing Himachal Pradesh. Uttar Pradesh.

West Bengal. Sikkim. Assam. Arunachal Pradesh. Meghalaya. Mizoram and Nagaland inhabits the temperate tasar silkworm. The food plants is oak plant. There are five species in temperate tasar.

1. *A. proylei.* Jolly
2. *A. roylei.* Moore
3. *A. pernyi.* Guerin-Meneville
4. *A. yamamai.* Guerin-Meneville
5. *A. polyphemus.* Cramer

The primary food plants are

1. *Quercus incana*

2. *Q. Serrata*
3. *Q. delabata*
4. *Q. himalayana*
5. *Q. semiserrata*

The morphological features of life stages and rearing aspects are similar to *A. mulitta.*

MUGA SILK

The golden yellow silk is secreted by multivoltine silkworm *A.assamensis*. distributed in Assam. The popular items made from this silk are 'dhoti'. 'chaddar'. 'chapkan'.'pugree' and 'mekhala'. Commercial reaing is practiced in Sibsagar. Lakhimpur. Nowgong. Darrang and other districts. If the larvae are fed on mejankori leaves:/ *A. assamensis* produces mejankori silk. This silk is very much admired for its durability. luster and creamy white shade. The muga reeling and weaving are done at Sualkuchi village.:The eggs are brownish measuring 2.8x2.5 mm in size and 9mg in weight. Newly hatched larva has prominent black intersegmental markings on yellowish body and brown head.

It measures 7 mm length and weights 7mg. The mature larvae measures 9X2cm in length and weighs 15 gr. In third instar the prothorasic hood markings consists of two prominent rectangular black marks which are replaced by semilunar dark brown markings in the remaining instars. The and flap carried rectangular black mark: which becomes U-shaped in IV instar and finally changes to a 'V':.

The pupa is copper brown measuring 3.2 x 1.8 cm and weighs about 5.7 gr. The cocoon is single shelled. light brown. oblong. closed. reelable and slightly flossy with a weak pendunclе. It is golden brown or glossy white measures 5.2 x 2.4 cm and weighs 6.3 gr. Female moths are longer: than males:. the wings are brown rarely with a pinkish tinge. Wing span in males is about 13 cm and 15 cm in females.:The polyphagous silkworm feeds on *Machilus bombycina* King (Som) and Litsaea monopetala. Pers.:. Other food plants include *L.citrata* Blume; *L. salicifolia.* Roxb. Ex.Wall; *Michelia oblonga.* Wall; *Cinnamomum obtusifolium.* Nees; *Symplocos grandiflora.* Wall.

The polyvoltine is reared outdoors except for indoor cocooning. Rearing is conducted round the year:. These crops are designated by local names based on rearing months (Jarua. Jethua. Aherua. Bhadiya Kotia). Kotia and Jethua are large scale crops for the production of reeling cocoons. other crops are stored for seed and stock.

Hatched larvae along with kharikas are tied to main trunk of the food plant. The larvae crossing the barrier are collected using bamboo sieve known as 'chaloni' which is hung on a leafy twig. Mature larvae are placed in a pit or bamboo basket. above which jail: is placed.

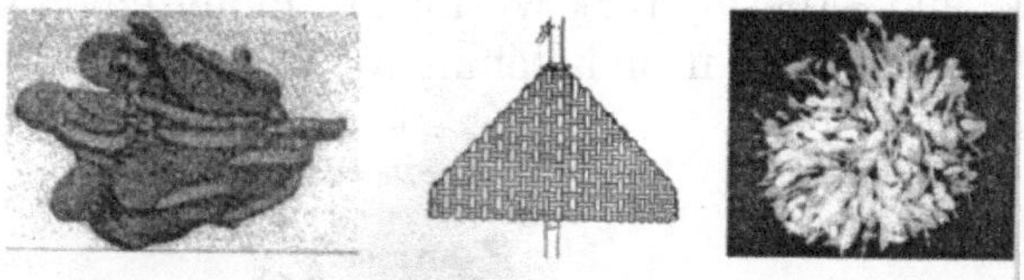

Worms on Branches of Food Plant Chaloni Cocoon Formation

Fig. Muga Rearing

ERI SILK

The white or brick – red eri silk: is produced by Philosamia ricini. B. a domesticated multivoltine silkworm. It is widespread in Assam and also practiced in Bihar. West Bengal.

Manipur. Orissa and Tripura. Among the non-mulberry varieties. eri has the disadvantage of higher production costs because it is made from domesticated silkworms.:The eggs are ovoid. candid white measure 1.5 x 1.0 mm and weight 6 mg. On hatching the larva is greenish yellow measuring about 5 x 1 mm and weigh 1.5 mg. the larval body colour changes gradually to pure yellow by the end of third day. From third instar onwards the body colour segregates into yellow. cream. green. blue or white.

Mature larvae measures about 7.0 x 1.5 cm and weighs 8 gr. Is translucent and covered with a white powdry substance. Spotted and unspotted larvae are found. The spots are of various types *i.e.* single. double. zebra and semizebra. The prothoracic hood of the first-instar larva has a black dorsal band. which splits up into a pair of crecent shaped markings in the second and third instar. These markings disappear at the fourth instar. The planta. anal flap and claspers are light yellow throughout the larval span. The tubercles. setae are also present. Pupa is obtect. adectious measures about 2.8 x 1.5 cm and weight about 2.6 gr.

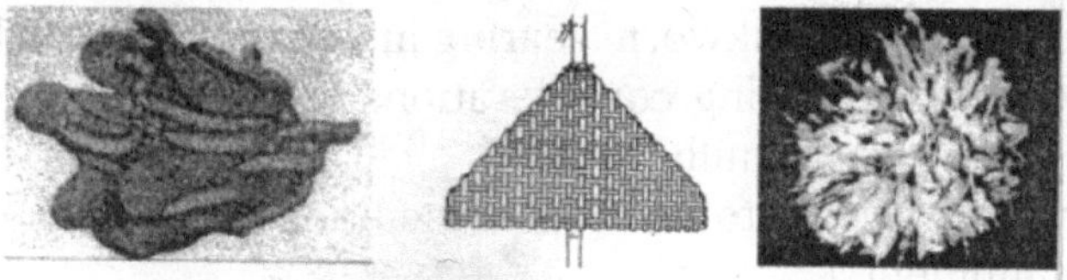

Worms on Branches of Food Plant Chaloni Cocoon Formation

Fig. Eri Life Cycle

The silkworm feeds mainly on *Ricinus communis:*. Other important secondary food plants are *Heteropanax fragrans*. Seem; *MManihot utilissima*. Phol; *Carica papaya*. L;*Evodia faxinifolia*. Hook; *Ailanthus excelsa*. Roxb. Rearing of these silkworms is entirely different from other wild silkworms. It requires a ideal rearing house and a rearing rack worms are mounted on split-bamboo tape:. Rearing of 100-125 layings from hatching to ripening requires about one metric ton of foliage.

ANAPHE SILK

Southern and Central Africa are the main source of this silkworm. It is distributed in intertropical regions of continental Africa (Nigeria. Uganda. Cameroon. the Congo. Togo). The important species used in the production of anaphe silk are *A. infracta*. Walsingham; *A. venata*. Butler; *A. panda*. Boisduval; *A. reticulate*. Walker; *A. carteri*. Walsingham; *A. moloneyi*. Druce and *A. ambrizia*. Butler The Anaphe are polyphagous. So far twenty two food plants are recorded (Albizzia fastigiata; Sterculia tragacantha; S.setigeri; S. rhinopetala etc.. are found in Nigeria and Bridelia micrantha; Cynometra alexandri; Triumfetta macrophylla are in Uganda).

The freshly laid eggs are soft and creamy white. but later become hard and grey measuring 0.97x0.62 mm. the larvae are green. Pairs of tubercles present on the thoracic and abdominal segments bear short brown setae. Long tufts of white hairs are dispersed over the body. The pupae are obtect adectious and dark brown. The wings of the moth are creamy white with a span of 4-6 cm. The anaphe silkworm is univoltine and undergoes diapause at the pupal stage. The silkworms from the cocoons communally the cocoon is large nestlike: weighs about 3.5 kg. generally 12-100 worms collectively spin cocoon for about 4 months. These cannot be reeled and looks like mulberry 7 cocoons.

REARING HOUSE

In temperate or sub tropical climates (Jammu and Kashmir. Punjab and Dehradun area of U.P) silkworm rearing is carried out twice a year: or three times: according to the growth and production of mulberry leaves. In tropical areas (Karnataka. West Bengal. Andhra Pradesh and Tamilnadu etc..) mulberry is full and flushy throughout the year favouring the famer to rear silkworms five to six times in a year./ the silkworm rearing in a farm house can be achived to the best. provided the following considerations are fulfilled *i.e.* conditions mulberry growth and yield of mulberry leaf. availability of laour for rearing silkworms. variety of silkworm to be reared. facilities for rearing silk worms *viz.* building and equipment.

CONDITION OF MULBERRY GROWTH AND AVAILABILITY OF MULBERRY LEAF

The mulberry plant come to "full flush" in about six months from the time of planting. The yield and quality of leaf depends on the agronomic practices followed for the cultivation of mulberry *i.e.* irrigation. application of manures and fertilizers. cultural operations. The size of the silkworm rearing depends on the availability of mulberry leaf yield. An hectare mulberry garden yields 20.000 to 25.000 kg of leaves per year under optimum agronomic practices. An hectare of irrigated mulberry yields about 30.000 kg of leaf in a year in Karnataka. W.B.. A.P. and Tamilnadu. In case of pit system under rainfed condition. the yield is about 10.000 kg per year. Therefore the leaf requirements are as follows.

No. of DFL's	Breed	Required Mulberry Leaf in Kg
100	MV x BV	700
100	BV x BV	800

On the basis of seasonal availability of leaves. the number of layings to be brushed is decided. However it is necessary to avoid shortages of leaf during rearing as they lead to poor crops and also failure of rearing.

LABOUR AVAILABILITY

Sericulture is labour intensive industry and there has been very little menchanisation of its operations. The most ideal silkworm rearing will be one Which is managed entirely by the family labour of the sericulturist. But practically it is not feasible in large scale economic rearing. Therefore the rearer should go in for hired labour at least in later stages of rearing. both for supplying mulberry leaf and mounting of ripe worms. Any shortage of labour or if labour costs are high. economic production costs will also increase. Agriculture and sericulture have to draw on the same surplus labour force in the rural parts. has to be ensured that the cocoon harvesting time does not clash with major agriculture operations in the village. Technologically developed sericultural countries like Japan. about three man hours are required to produce one kg. of cocoons. In India it would be safe to assume about four to five hours. where traditional methods are followed and man hours can be reduced if rational methods are followed.

VARIETY OF SILKWORM TO BE REARED

Depending on the voltinism the silkworms are broadly devided into univoltine. bivoltine which give one and two crops respectively in a year. Where as multivoltine worms give four or more number of crops in a year. In Karnataka "Mysore" race is very hard and gives greenish yellow cocoons with poor built. It is grown in seed areas for raising local race seed cocoons. The female parent of this is taken for hybrid seed for the preparation of reeling cocoons and to get commercial silk. In West Bengal another multivoltine race called "Nistari" which gives deep yellow cocoons are popular. However in Jammu and Kahsmir bivoltine hydrids are rared. The productivity of uni/bivoltines is better than that of multivoltines though they require more favourable temperature and humidity condition: Bivoltine pure or local race cocoons fetch premium price than reeling cocoons. Thus large number of farmers rear commercial hybrids *i.e.* female local X male bivoltine or female bivoltine X male bivoltine. New breeds released by Central Silk Board (CSB). *i.e.* Nandi a bivoltine hybrid is more popular in Karnataka.

SITE SELECTION

Silkworms are more sensitive to environmental conditions especially to temperature and himidity. So selection of rearing house should be planned and constructed to maintain optimum environmental conditions and to get good quality cocoons. In Indian conditions the farmer is financially handicapped to built a new rearing house. But proceeds to rear in the suitable part of his existing house. However. the site of rearing house should not be water-logged. posses provision for fresh pure air. without being exposed to direct heat of the sun or draughts. If the surroundings are damp. it should be cleared by making drains or basement hight should be increased. Humidity can be reduced by opening suitable windows. allowing free air. If the moisture of air is due to river or lake situated near the house. the direct current of air can be avoided by raising bushes or trees or walls or wind breaks. If there is any stagnation of air around the rearing room. it is cleared by cutting thick vegetation. On the otherhand violent draughts of wind. heat from the sun. cold from chill winds at night can be prevented by raising tress and suitable shelter.

Orientation and Construction of rearing house

Orientation of the room should be such that the interior is protected from the direct afternoon sunlight. as it is harmful to worms. The best orientation of rearing room will be north and south. the room facing north with the door open to cool breezes. with a window on the south side to aid ventilation and with the sides of the room east and west. If this direction is not possible the next best would be a north – west face. and south east back. Otherwise if a farmer is able to built a new house east is better than west face. East facing should be all right in the morning. but as the day increases the room warmup and cool at night with the cold night air. On the otherhand afternoon sun will make the room hotter and hotter. leading to drying up of leaves. West facing would be the least facing should invariable posses broad verandah or trees planted to cast a shadow. Inside the rearing room mulberry leaves and silkworms should not be kept on western side as it is hot in the afternoon. Opposite tot eh door. a window should always be provided to induce ventilation. While building a new house. the width of the room should be smaller than its length or frontage and should not exceed 15 to 18 ft. A broad room leads to inequalities of temperature. On the other hand a small room leads to stagnation of air. for which windows and ventilators should be provided to admit free passage of air. Verandahs 4-6 ft. broad are provided at least on that side which is most exposed.

In any construction. roof should be of non-conducting material and amply provided with ventilators. For tropical conditions mud wall and thatched roof or a country tilled roof is always ideal. But corrugated zinc sheet roof is the worst of all. For any kind of rearing house high roof is better than a low roof and a false ceiling is desirable. It is necessary to provide doors. windows ventilators etc. with fly proof mesh. However. all the above perfections which could be adopted by rich persons who intend to construct a new rearing room. But in India sericulture is mainly confined to marginal farmers who can not spare a new rearing house. In any case the sericulturist must remember the needs and necessities for the better growth of silkworms and working knowledge. which enables him to make the best of his limited sources. The following are some of the important guidelines to be remembered:

AVOID:

1. Dampness
2. Stagnation of air
3. Direct and too strong draught of air
4. Exposure to bright sunlight and radiation
5. Narrow rearing space.

ENSURE:

1. Adequate temperature and humidity
2. Good ventilation
3. To provide two shutters to the door. Further windows and doors are fitted with wire mesh.
4. Rat proof building by providing a projecting edge all round at plinth level.
5. Movable stairs.

CENTRAL SILK BOARD (CSB) MODEL

It is a rat-proof building with aledge all around to prevent rats entry into the building. The building has a verandah all around and glass windows and doors to provide good ventilation and light. The ceiling of the rearing house is generally made of wood. if made of concrete or tiles a false ceiling must be constructed. Ventilators must be installed to ensure free circulation of air. The rearing house is partitioned into four convenient rooms in one of which high temperature and humidity are maintained for the young age silkworms rearing. The rooms are provided with an adequate number of windows and doors top ensure good ventilation for rearing late age silkworms. The rearing houses should be located in such a manner so as to maintain as possible ideal temperature and humidity conditions inside the rearing rooms. In temperature and sub-tropical regions they should be constructed in northsouth direction so that maximum sunlight and heat is available to adequate warm up of the rearing rooms. In tropical regions. however the building should be sited east-west so that too hot. direct sunlight is avoided and cool room temperature are maintained. Recently there has been a tendency to use improvised rearing sheds with roof and side walls made of plastic sheets or taurpaulins which prevent direct draughts of cold air entering from outside the sheds. This type of shed is usually used to shoot or floor rearings. In certain countries such as India. where mostly shelf rearing is practiced there are no separate rearing houses as such. but the rearing is done in the dwelling houses only. Similarly in Jammu and Kashmir of India. where most floor rearing is practiced in the rearers own houses. The following are different rearing houses suitable for different regions.

1. CSB model rearing house
2. Two rack capacity rearing room
3. Double wall and mangalore tile roof model
4. Mudwall and country tile roof model
5. Mudwall and mangalore tile roof model
6. Mudwall and thatched rood model
7. Bamboo mat wall and thatched roof model
8. Plankwall and thached roof model

In the end. we can say:

- The rearer should think about favourable conditions for mulberry growth labour availability and decide about variety of silkworms to be reared.
- There are multovoltine. bivoltine varieties which are better suitable Indian conditions.
- Silkworm rearing requires a suitable rearing house and favourable environmental conditions.
- Rearing room surroundings should be healthy and neat.
- Rearing room should not allow direct sunlight.
- While building a new house. the width of the room should be smallest than its length.
- Rearing room roof should be of non-conducting material and roomis:provided with ventilators.
- Mud wall and thatched roof or a country tiled roof is always ideal for tropical conditions.
- Besides CSB model. different types of rearing houses: are available.

REARING EQUIPMENTS

The size of the silkworm rearing and number of equipment required depends on the availability of mulberry leaf. An hectare mulberry garden yields 20.000 to 25.000 kg of leaves per year. An hectare of irrigated mulberry yields about 30.000 kg of leaf in a year in south and west areas of India. In case of pit system under rainfed condition. the yield is about 10.000 kg per year. Keeping in view of rearing conditions. labour availability the variety of silkworm to be reared is decided. The rearing equipment especially rearing stands. leaf chamber are to be placed at proper place. The space available in the rearing room has to be completely used for rearing stands and trays so that maximum possible rearing can be achieved.

EQUIPMENTS AND USES

Rearing stands

These are frames on which rearing trays containing the silkworms are placed. These stands are made up of wood or bamboo. The bamboo rearing stands are very much popular in Karnataka and West Bengal. It measures 2.25m. height. 1.5m. length and 0.65m. width. It should have ten to twelve tiers spaced at a distance of 0.15m. between the two tiers. Six stands are needed for each rearing room:.

Ant Wells

Ant are serious pests of silkworms. They are to be prevented before they crawl on the stand. It is possible by resting the rearing stand legs on ant wells. The simplest ant well is an enameled plate:. Cement ant wells (21x21x8cm size with a groove of 4 cm) can also be used. The groove is filled with water to prevent ants:. on the otherhand a piece of cloth dipped in kerosene can be placed around the legs of rearing stand. Dusting of BHC also serves the purpose.

Rearing Trays

Rearing trays are portable equipment for keeping silkworms during rearing. There are many kinds of trays differing from one another in materials. shape and size. The selection of the material should not be costly as sericulture is poor man's industry and luxurious equipment are quite out of place. however the material should not be very flimsy and perishabel as it would become costly in long run. In India bamboo round trays are popular as they are light and easy to handle and carry from place to place. It is cheap and can be prepared in every village. It measures 138 cm diameter with a depth of 6.5 cm:. Wooden trays: are used for rearing young stage silkworms. Two trays are required till the end of second instar for rearing 100 DFLs. These wooden trays are accommodated in bench type young silkworm stand:. The trays are piled

Paraffin Paper

This is a thick craft paper coated with paraffin wax with a melting point of 550C. it is used in modern methods of chawki worm rearing. It is to cover the silkworms bed to prevent leaf withering and to maintain required humidity.

Foam Rubber Strips

For modern method of rearing long foam rubber strips: are essential. These are dipped in water and kept around the chawki worm bed to maintain optimum humidity. As a substitute. news paper folded into convenient strips are also used.

Chopsticks

This bamboo stick is 17.5 to 22 cm long. thin in girth and tapering at one end. These are used for pocking the young larvas. This ensures hygienic handling of delicate young worms:.

Feather

A white quill: feather is most essential in the rearing room. It is used for brushing newly hatched worms and also for changing beds in the early stages. It is the most convenient and safe for brushing:.

Chopping Board. knives and mats

The mulberry leaf are chopped according to the age of the worm before feeding. This equipment is essential for chopping the leaf in a proper way. The board is made up of soft wood (92X92 cm with 7.6 cm thickness). Each rearing house needs two knives (small one for chawki rearing. big one for late age rearing). The knife is 0.3 to 0.5 m long blade with 4 to 8 cm breadth:. Leaf chopping is carried on the mats to avoid dust. dirt and helps to collect the chopped leaves.

Leaf Chamber

The rearer can not go to field to get the leaf for every feeding. And it is also not advisable to cut the leaves when it is too hot or raining. The required leaf is therefore plucked by hired labour in the early hours or late in the afternoon. Thus it is necessary to preserve the nutrients of mulberry leaf till feeding. It can be achieved by preserving the leaf chamber/eartherned pots/ refrigerators. A simple leaf chamber is made up to wooden strips (152 X 76 X 76 cm size;strip size is 7.5 cm fitted with 7.5 cm spacing). The chamber is covered on all sides with gunny cloth which is made wet. During summer season and dry days water is sprayed periodically on the gunny cloth:. The leaf can also be preserved in earthened pots:. the rearer should remember that heaping of leaves in a corner is worst. Because the leaves rapidly wither owing to evaporation especially in dry season. On the other hand temperature inside the heap raises favouring the fermentation in the leaves. Besides this there is every danger of the leaves getting infected with diseased germs. dust especially during cleaning. bed changing etc. thus it is necessary to preserve the leaf in leaf chamber

Cleaning Nets

During the course of feeding certain amount of mulberry leaf is left behind in the tray. This dried leaf together with litter of the worms accumulates in the rearing beds. owing is increase of bed size. Further is also adds to increase temperature and harmful gases in the bed. Thus the waste leaf as well as litter has to be periodically removed using nets. basing on the age of the worms. The nets are woven using cotton or nylon thread. The mesh should be square and of the same length of the silkworm. The rearer should possess small:. medium: and large: nets: for rearing chawki and late age worms.

Mountages

It is used to enable the ripe worm to spin cocoon. The most. common form of mountage in India is "Chandrika". It is a rectangular bamboo mat on which a spiral bamboo tape is tied. The chandrika measures 1.8 m X 1.2 m. The tape is about 4-5 cm. broad and space between the spirals is about 4-5 cm. it became more popular than any other mountage because of the following advantages:.

a. Easily manufactured in the villages by bamboo weavers.
b. Can be stored easily.
c. It provides easy passage of air for quick drying of excreta of spinning worms and avoids staining.
d. Easy to transport.
e. Easy to disinfect.
d. Low cost and light weight.

Besides chandrika there are other types of mountages which are in practice *i.e.*. zig-zag mountage. centipede mountage. rotary mountage etc. Zig-zag mountage frame is made of plastic wire or sedge straw/mulberry branches/ bamboo. usually consists of about 40 corners of about 6 cm each. and each set can mount 250-300 worms.:. In India straw mountages in Assam. mulberry branch mountages in Jammu and Kashmir are in practice. Recently P.V.C. mountages are also used. It has the same advantages like chandrika besides reducing the labour expenses.

Rotary mountage has pieces of cardboard to form 13 rows. consisting of 12 sections and each and amounting to 156 sections. Ten pieces are put into frame as a set:. when this frame is hung up with wire holding at by ends. the frame can be turned around two axes. This frame produces fewer spoiled cocoons and raises the reelability of cocoons. Good cocoon percentage is more than 80. It is the best type and suitable for large scale silkworm rearing because it does not only lead to increased cocoon quality but also saves labour in mounting and harvesting.

Centipede mountage is made using plastic or straw material. It provides for self mounting similar to rotary type. But care should be taken to avoid a high density of worms on the mountage because it may cause a higher incidence of bad quality cocoons *i.e.* double and deformed cocoons. It is easy to transport and low cost:.

Foot Cleaning Tray

Silkworms are very sensitive to dust and micro organisms which may cause damage. Thus they are prevented prior to their entry into rearing room along with foot of rearer. This can be achieved by keeping foot cleaning tray at the entrance of the room. It measures 1.1m X 0.5m X 0.1 m with tin bottom. A pad of gunny cloth soaked in 2per cent formalin in kept in the tray. Thus the persons entering into the rearing room disinfect their foot:.

Basin and Stand

While entering into the rearing room it is advisable to disinfect hands before handling the worms. Thus basin. stand: is required to keep disinfecting solution:.

Feeding Stand

At the time of feeding the worms. wooden feeding stands are required for keeping the trays.:. It can be folded:.

Other Appliances

For reading temperature and humidity in the rearing room dry wet thermometer. hygrometer. humidity chart are essential. During winter especially in Kashmir. West Bengal it is necessary to heat up the rearing room to increase the temperature. Thus an electric. charcoal stove is essential. On the otherhand it is necessary for fumigation process also. Sprayer is required for disinfecting the rearing room. equipment and also spraying water on the leaf chamber. Leaf baskets to carry leaves from the mulberry field to rearing house. litter basket to collect waste leafage. litter etc; disinfection mask for workers remaining in the rearing room for longer period for effective disinfection; black box/paper/cloth for black boxing of eggs; humidifier for increasing humidity are also required for rearing room. For measuring chemicals/formalin and to weigh the leaf for assessment of growth a rough balance. measuring cylinder. buckets are required. Other items like gunny cloth. uzi proof net are also required.

CHEMICALS AND USES

For successful rearing of silkworm maintenance of hygienic conditions are necessary. To prevent and control the micro organisms disinfectants like formaldehyde. paraformaldehyde. bleaching powder. sodium hypochloride. slaked lime powder are necessary. Silworms are not so exception for diseases. Thus as a preparatory step Reshamket Oushadh (RKO). uzicide. China clay. dimilin and fungicides like Dithane M-45 or Captan are kept along with other chemicals.

- Summary rearing size and equipment number required depends on the availability of mulberry leaf.

- Bamboo rearing stands are popular which have 10-12 tiers.
- Ant wells are necessary to prevent ants.
- Paraffin trays made up of wood. bamboo are required. Bamboo trays are cheap and light.
- Paraffin Paper. foam rubber strips are essential to maintain proper tem perature and humidity during chawki rearing.
- Chop sticks ensure hygienic handing of delicate young silkworms.
- White feathers are necessary for brushing.
- Silkworms are fed with finely chopped mulberry leaf. This function requires necessary equipments.

PREPARATION FOR REARING

Like any other animals silkworms are also susceptible to a number of diseases. Some microorganisms like bacteria. virus. fungi. protozoan parasites attack the silkworms readily and spread the disease quickly. The optimum temperature and humidity for the silkworm are suitable for the growth of such germs and insects etc. therefore maintenance of hygienic conditions in and around the rearing room before starting rearing of silkworms enables to avoid the disease causing organisms.

CLEANING

This process has to be done prior to disinfection. After selection of rearing house before starting rearing of silkworms the rearing room and rearing equipment are to be cleaned. All the dumped materials. waste materials. dirt stones etc. in and around the rearing room are removed and cleaned using broom stick. The surroundings of the rearing house must be kept neat. If there is any stagnation of water around or near the rearing room. it becomes a reservoir for microorganisms to develop.

Therefore it is necessary to clear the stagant water. After cleaning the dust and dirt around the rearing room. water mixed with cow dung is sprinkled. The ceiling and walls of the rearing room has to be cleaned. All the crevices and holes in the rearing room should be closed using clay mixed with cow dung. The rearing room and appliances are washed thoroughly with water and then dried in sun:.

PREPARATION OF DISINFECTANTS

For preparing required strength of formalin the following methods are used.

First Method:

$$\frac{\text{Required strength}}{\text{fo Formalin}} = \frac{\text{strength of original formalin - strength of formalin required}}{\text{Strength of formalin required}}$$

The required strength of formalin is prepared depending on the surface area of rearing room. A surface area of 10' x 10' requires one Litre disinfecting solution. Therefore a room size of 10' x 1-' requires roughly six Litres of solution. For disinfecting rearing equipment additional two Litres of solution is required. From the above formula we get the ratio of water to be added to one part of commercial formalin to get required strength. Commercial formaldehyde available in the market is 38-40per cent. Disinfection by formalin can be strengthened when wetting agent like detergent: added in 0.2per cent concentrations. These agents improve spreading and penpetration of disinfectant.

Model Problem

Calculate the required 2per cent formalin solution and also the required amount of commercial formalin for dinsinfection of rearing house of 6X9 m with 4 mts. Height with terraced roof.

Solution

The required of 2per cent formalin solution for disinfection of rearing house of 100 sq. m. area is about 5.73 litre. This is calculated as

1. Floor area = length x breadth
2. Area of 2 walls = length x height of each wall x 2
3. Area of 2 other walls = breadth x height of each wall x 2
4. Roof of terrace = length x breadth.

To disinfect 6 x 9 m with height of 4 mts size room. the required 2per cent formalin is calculated.

Floor area = 6 x 9 mts = 54 mts 2 or (20' x 30') = 600 sq.ft.

Area of 2 walls = 6x4x2 mts = 48mts2 or (20' x 10.3'x 2') = 412 sq.ft.

Area of 2 walls = 9x4x2mts = 72mts2 or (30'x10.3'x2'_ = 618 sq.ft.

Area of roof = 6x9 mts = 54mts2 or (20'x30') = 600sq.ft.

Total area	**228 mt.sq.ft. 2230 sq.ft.**

$$\text{Requirement of 2\% formalin solution} = \frac{228 \times 5.73}{100} = 13.06 \text{ liters}$$

For disinfecting the rearing room 13.06 Litres of 2per cent formalin is required. Additionally equal amount of solution is required for spraying rearing equipments..:Total requirements of 2per cent formalin is 26 liters.

$$\frac{40 - 2}{2} = \frac{38}{2} = 19 \text{ liters}$$

One Litre of commercial formalin is to be added to 19 liters of water to get 20 liters of 2per cent formalin. to get 26 liters of 2per cent formalin.

$$\frac{26}{20} = \frac{13}{10} = 1.3 \text{ liters}$$

About 1.3 liters of commercial formaldehyde is required to disinfect rearing equipment and rearing house: which can accommodate 250 - 300 DFL's.

Second Method

$$\text{Quantity of formalin required} = \frac{\text{required concentration X required quantity of solution}}{\text{available concentration of commercial formaldehyde}}$$

Solution

$$\frac{2X100}{40} = \frac{2000}{40} = 50\text{cc of commercial formalin}$$

for disinfecting rearing room: and rearing equipment 26 liters of 2per cent of formalin is required. To calculate quanity of formalin requires is

$$\frac{2x26}{40} = \frac{52}{40} = \frac{13}{10} = 1.3\text{ liters}$$

1.3 liters of commercial formalin is required

DISINFECTION

The distruction and extermination of disease causing germs is known as disinfection. The most effective and simple method of disinfection is by using various chemicals and disinfectants. For the selection of disinfectant the follow3ing points have to be considered.

- Effectiveness against different pathogens.
- Simplicity. ease and swiftness of its application
- Harmless nature to human beings and domestic animals
- Easy availability
- Cheap cost.
- Harmless to buildings equipments.

The commonly used disinfections which are effective against various pathogens in sericulture industry are.

1. Formalin
2. Paraformaldehyde
3. Bleaching powder
4. Sodium hypochloride
5. Slaked lime powder etc.

However the effectiveness of disinfectants depends on the concentration. duration of contact ambient temperature and humidity. The rearing room and equipments have to be prepared in advance for ensuring effective and easy disinfection. This involves;

1. Collection of dead larvae. pupae. cocoons used paper bits. other debris etc. and burning them.

2. Taking out of rearing equipments and washing with either 5per cent bleaching powder solution or 0.5per cent sodium hypochlorite solution.
3. Sorubbing and washing the floor and walls of the rearing room with 5per cent bleaching powder 0.5per cent sodium hypochlorite solution.
4. Drying of equipments in the sun.
5. Sealing of cracks and crevices in the rearing room.
6. Making the room air tight by closing windows and doors.

Rearing stands and chandrikas are to be cleared of dead larvae. cocoons and floss before washing. The equipments and the rearing rooms should not be washed with only water as disease causing pathogens. get washed out and are deposited outside the rearing room. Thus they may be carried into the rearing room again and become source of infection. Rearing rooms and instruments (except for those made up of cloth or paper) are washed with water carefully before they are disinfected. After washing they are dried in the sun. sun drying is a kind of disinfection. The articles made up of paper or cloth are dried in the sun sufficiently. During sterilization if 1/ 500 high power bleaching powder is added to water. increases the sterilizing power.

Disinfecting methods

Disinfecting methods are two kinds-physical and chemical. In physical method steam or boiling water is used while in the later method various kinds of chemicals are spread. Physical disinfection is only for wood or metal made equipments. It is not suitable for a wide area or for the rearing room.

Disinfection of rearing equipments.

The rearing equipments such as rearing trays. feeding stands. basins and wells etc.. which are smaller in size and easy to handle are first disinfected. These are dipped din a tank containing 0.5per cent sodium hypochlorite or 5per cent bleaching powder solution for at least 10 minutes followed by sun drying. Cleaning nets. gunny cloth. towels. foam pads etc/ are soaked in 2per cent formaldehyde solution. The bigger equipments such as stands. leaf chamber. chandrikas etc.. are disinfected by spraying either 2per cent formalin or 5per cent bleaching powder solution.

Articles which can not be soaked or sprayed with a liquid disinfectant are disinfected by fumigation. In India it is a common practice among the rearers to smear the bamboo trays with cow dung to increase durability. It is also suggested that cow dung is made to slurry with 5per cent bleaching powder or 2per cent formalin and smeared. This improves the durability as well as disinfection simultaneously. An alternate method of the above is to use a paste made out of soaked menthol seeds and news papers. The chandreka's having floss after cocoon harvest wil be disinfected through burning of floss using a burner.

Disinfection of rearing house

Before disinfection the rearing house is made aitight abd 2per cent formalin solution is sprayed using sprayer. All the rearing equipments. walls and floor are thoroughly made wet by drenching spray. The spray should be uniform with sufficient pressure. For this a power guitor rocker sprayer is suitable. A strong spray boom and a solid cone spray nozzle may be used. Spraying of formalin solution makes metallic machines and equipments rust. So care should be taken not to spray the formalin directly on them. Ideal room temperature should be maintained at 250C during disinfection for quick diffusion of formaldehyde gas. When the temperature after disinfection is low. the sterilizing power falls. If the temperature is low. room is to be heated. The proper time for disinfection is around 11 A.M.. As the day increases the raised temperature in the afternoon helps in better disinfection. After spraying the room should be kept closed for atleast a day. More efficacy of formalin disinfection against virus. a mixture of 2per cent slaked lime is advisable:. Formalin causes irritation in nasal mucosa and eyes. further the skin hardens.

Therefore care should be taken during spraying by keeping nasal mask and gloveses. Chlorine compounds such as bleaching powder or sodium hypochlorite solution can be used as disinfactants. Bleaching powder is dissolved in water to get 5per cent solution. After shaking well supernatant solution is used for spraying. When bleaching powder is used as disinfectant it is always better to chose freshly prepared or properly packed power.

The chlorine available in bleaching powder is unstable thus stored powder looses its disinfection capacity. This powder is good against cytoplasmic polyhedrosis virus (CPV). nuclear polyhderosis virus (NPV). flacherie fungus causing muscardine and pebrine spores. Disinfection in the gaseous form is called fumigation. However this process of disinfection is effective only in air tight rearing houses. This method is convenient and easy to disinfect rearing houses and equipment at a time. For fumigation process formalin can be used. The quantity of original formalin solution required for disinfection is calculated according to the room size. Then the quantity is diluted – times.

The solution is allowed to evaporate in a pan. by heating from beneath using electric or charcoal stove. Slowly the formalin gas comes out. This process should be conducted in the air tight rearing room. Care should be taken that the chemical does not catch fire. which will nullify the disinfecting effect. In any case it is necessary to fumigate for 4-5 hrs. the temperature of 200C or higher and 70 percent or above humidity is preferable for disinfection. Disinfection of metallic instruments can be better achieved by fumigation process.

Paraformaldehyde can also be used for fumigation process as it releases formaldehyde fumes on heating. The quantity of paraformaldehyde required to fumigatean area of 10 cubic meters is about 60 grams. Rearing house should be closed after disinfection at least for one day. All the doors. windows are opened 1-2 days before use of rearing house to allow free circulation of fresh air.

MAINTENANCE OF HYGENIC CONDITIONS

Maintenance of good hygienic conditions during rearing results in better crop production. As this process not only helps in preventing the entry of disease causing pathogens from out side. but also checks the spread of various diseases. During the process of rearing only a few members are allowed into the rearing room. While entering into the rearing house. they are advised to wipe their feet on a foot cleaning tray and wash their hands in wash basin. Disinfectants like formalin. detol or cresolor soap is added to the water n wash basin. Bleaching powder or a freshly prepared slaked lime is sprinkled on a gunny cloth placed on the foot mat at the entrance of the rearing house. Freshly prepared lime is also sprinkled all round the rearing room. On the other hand diseased and dead worms collected during rearing should be put in lime vats or containers.

The concentration of formalin solution in the basin stand should not increase one percent. The silkworm litter compost pit should be far away from the rearing house. While bed cleaning care must be taken to prevent dropping of litter on the floor of rearing house. Immediately after cleaning. the room should be sweeped and the rearer should wash hands and feet. Holes and crevices of rearing room should be closed to check the entry and accumulation of litter in them along with disease causing organisms. Therefore cleaning of rearing bed should be done preferably with nets. If the incidence of disease is more the worms are immediately shifted in to disinfected trays. Rearing equipments should not be borrowed of lènd to prevent the transfer of pathogens from rearing house to the other. The rearing articles such as paraffin paper. old news paper are used for only one rearing. In any case overlapping of rearings should be avoided to maintain proper hygienic conditions and also to get good cocoon production.

- Maintenance of proper hygienic conditions is an important aspect in silk worm rearing.
- Cleaning of dust and dirt in and around the rearing room is ensured. All the crevises. holes and room. equipments are cleaned properly before disinfection.
- Disinfection includes elimination of various harmful microorganisms using various chemicals *i.e.* formalin. bleaching powder etc.
- The equipments have to be washed with water. dried and disinfected.
- There are two methods of disinfections – physical and chemical.
- For disinfection 5per cent bleaching powder/ 2per cent formalin/0.5per cent sodium hypochlarite is used.
- During disinfection 250C temperature is maintained for quick diffusion of formalin gas. Room should be closed one day after disinfection.

- The required strength and amount of disinfectant is calculated according to the room area.
- Fumigation method of disinfection is effective in air tight rearing houses. Fumigation is continued for 4-5 hours. where disinfecting solution is warmed using stove.
- Maintenance of hygienic conditions checks the spread of diseases.

ENVIRONMENTAL CONDITIONS

Silkworms a ether domesticated insects for many centuries. as silk played an important role in the economic life of man. The silkworms are by nature quite delicate and very sensitive to environmental conditions. Silkworm rearing is aimed to produce good quality cocoons of very high silk content utilizing the available optimum environmental conditions.

The ecological factors chiefly temperature. humidity. light and air during rearing show a significant effect on the growth of larvae and finally cocoon crop quality. However. the other factors such as quality and quantity of leaf supply and the techniques of rearing adopted. Such as feeding. cleaning. spacing etc.. also influence the rearing activity.

The silkworm Behaviour in relation to various environmental conditions varies with different stages of growth. Bad environment affects the growth of the worm.

There by reducing the quality and quantity of silk as growth of the worm is directly correlated to the development of the silk gland. So based on the above responses. suitable environmental conditions have to be adopted for obtaining the best results. The importance of various environment at conditions and their effects on various aspects of rearing and silkworm development in particular are detailed in this chapter.

TEMPERATURE

Silkworms are poikilotherms/cold blooded animals. thus they change their body temperature according to the environmental temperature. Temperature has direct effect on the various physiological activities of living organisms. Like other animals: the physiology of silkworm *viz.* metabolic rate. activity of enzymes. nutrients conversion. digestion. assimilation. excretion. nervous stimulations. hormonal actions are influenced by environmental temperature.

The silkworms are capable of growing in temperature ranging from 150C to 400C. But from the physiological point of view the ideal temperature ranges from 200C to300C. The silkworm rearing is aimed to produce good quality cocoons for which an ideal and desirable temperature ranges from 230C to 280C for getting maximum productivity. The effect of temperature on the physiology of silkworm during rearing can be divided into the groups which are as follows.

- The temperature. which is harmless to the growth of silkworms.

- The temperature. which is favourable for the healthy growth of silkworms.

The temperature. which is favourable for making good quality cocoons. With the rise in temperature. the metabolic activities of the worm are increased while they are slackened when the temperature is low. Therefore at higher temperature the growth of the larvae is fast and consequently the duration of larvae period is shortened. On the other hand growth becomes slow at low temperature and it favours to prolong the larval duration. Therefore. to keep the temperature range with in the limit is absolutely essential for rearing silkworms.

The optimum temperature for normal growth in silkworms is between 200C and 280C. Temperature above 300C positively harmful to worms. If the temperature is below 200C. the growth is considerably retarded owing to very low rate of physiological function. This temperature problem especially in the early stages results to make the worms weak and susceptible to diseases. The chawki worms are comparatively stronger and more resistant to humidity temperature than late age worms.

Thus the young worms feed very actively and grow vey vigorous under high temperature and humid conditions. Such resistant and vigorous worms can feed and thrive better even under adverse conditions at later instars. Further. the resistance attained by early instar larvae improves the survival rate in various stages of development and finally top get good quality of cocoons. As it is said earlier. that the temperature has a direct correlation with the growth of silkworms. wide fluctuations are avoided as they are harmful. The maximum temperature for raring silkworms in different instars are as follows.

Age of worms		Optimum Temperature (°c)
I	---	26 - 28
II	---	26 - 28
III	---	24 - 26
IV	---	24 - 25
V	---	23 - 24

In case silkworms are reared in such a temperature in respective stadium. it is necessary that they are fed with rich nutrient sufficient leaves. So that maximum possible growth can be achieved. If insufficient amount of leaf is supplied expecting sufficient nutrition. it would be better to lower the rearing temperature at least to some extent. Moreover. regarding the rearing temperature in the former period of the stadium is higher and it is lower in later period of stadium. The rearing room temperature can be recorded using a thermometer.

HUMIDITY

The role of the rearing humidity upon the growth and health of silkworm is similar to that or rearing temperature. That is high humidity makes to decrease the length of growing period of silkworms by accelerating the activity of physiological functions. The pH value of the blood is remarkably lower at high humidity: than at low humidity: condition. Expiration of CO2 increases with rise in humidity. On the contrary. low humidity makes to prolong the length of growing period of larva. The combined effect of both temperature and humidity largely determines the satisfactory growth of silkworms and production of quality cocoons. In this way both temperature and humidity are complimentary to each another. The role of humidity is both direct and indirectly. It directly involves to influence the physiology of silkworm larva while indirectly influences the rate of withering of leaves in the larval beds. Therefore suitable rearing must be determined while considering the following points.

- The optimum humidity for the growth of silkworm is about 75per cent relative humidity. The early instars are resistant to high humidity with relatively little or no effect of change in the moisture. On the contrary the late age instars are weak against high humidity. suffering badly from the damage of humidity.
- In an extremely desiccated rearing room. silkworms cannot eat mulberry leaves owing to the withering of leaves. Then they become malnutritious especially this problem occurs in the early instars.
- If the air of rearing room is too moist. it becomes favourable to grow the pathogenic microbes thus silkworms are apt to suffer from diseases.

Considering the above factors. different humidity conditions for different instars should be provided as follows.

Age of worms		humidity%
I	---	85
II	---	85
III	---	80
IV	---	75
V	---	70

For knowing the humidity percent of rearing room a hygrometer/wet and dry thermometer and humidity chart are required. Keeping in view of the above considerations it is advisable that young silkworms are reared in a moist condition by covered box rearing. Because of high rearing temperature mulberry leaves are liable to wither and increase of feeding frequency is not advisable. On the other hand late age larva are reared in dried condition. Because the rearing room is apt to be wet with mulberry shoots with leaves. This realizes large amount of mulberry leaf supply. withering becomes slow. which introduces a wet climate in the rearing room. Moreover late age worms are weak to high humidity. At a humidity of 90 percent or higher. if the temperature is kept at 200C – 280C. the worms can grow without being affected.

AIR

Silkworms breathe through eighteen spiracles on both sides of the body. supplying the blood with oxygen through the tracheae. These tracheae distribute the oxygen to the body and at the same time unnecessary substances *i.e.*. CO2 or water produced in the body are exhaled through tracheae. Therefore it is need less to say that the fresh air is necessary for the life of silkworms. The air of rearing room is contaminated by carbon dioxide. sulfuric acid gas. carbon monoxide. ammonia. formaldehyde gas etc.. produced by the breath of man. silkworms or mulberry leaves. the fermentation of litter. burning of coal.

These gases are injurious to silkworms. When the gases increase in the beds and rearing room beyond the tolerance limits the worms become sluggish and do not feed.. The safe limit for silkworm rearing is 1 to 2per cent of CO2. 1per cent of formaldehyde gas. 0.02per cent of sulphur dioxide and 0.1per cent of ammonia in the air of rearing room. Young worms are less resistant to taxic gas though comparatively resistant to CO2. the production of these gases are far less in early instars than in late age instars. The effect of sulfur dioxide cause prothetely in the cocoons. and makes the cocoon quality worse and worsening the reliability of cocoon filaments. Ammonia gas: in the room makes the sericin insoluble while reeling.

LIGHT

There is no correlation between light and survival rate of silkworms. However. silkworms are photosensitive and generally tends to crawl towards dimlight. They do not like either strong light or complete darkness. to produce heavier cocoons than those of silkworms reared ion dark condition. The appetite of silkworm is more in a light place than in a dark place. The larvae come up to the upper surface of the bed more quickly under light condition than in a dark condition. However it is not advisable to rear either in complete light or complete darkness. Silkworms are fondof dim light of 15 to 20 lux and avoid strong and darkness. The late age worms thrive better in 16hrs. light and 8hrs. dark period. Light helps in uniform distribution of larvae in rearing bed. They are crowded and distributed in several layers in dark condition compared to thin and even distribution in light condition. Photo period influences on the early instars on the type of eggs produced by the resulting moth.

REGULATION OF ENVIRONMENTAL CONDITIONS

Regulation of Temperature

The temperature varie from season to season. Generally the rearing practice starts in the month of September depending on the leaf availability. The room temperature is considerably low during winter and rainy seasons. thus it becomes a hurdle for rearing. For providing required optimum temperature the room has to be heated with electric heater or charcoal fires. Electric heaters are however costly and a marginal farmer is not excepted to get that. But it is the best since it does not emit any smoke. When electricity is costly or not available. charcoal stove can be conveniently used. Care should be taken while using charcoal stove. as it should not emit smoke. It is better that the live clinders are covered with a layer of ash for better. more regulated and uniform dissemination of heat. In any case open fire using firewood as fuel should be avoided as it induces to generate obnoxious gases in the rearing room which inturn become harmful to worms. Besides this the doors and windows are kept closed during nights to keep out the cold and later in the day. as the outside temperature goes up. they should be opened to allow warm air to get in. This is another better way of temperature regulation without any expenses. In general. among the states Andhra Pradesh. Karnataka and West Bengal the temperature is often above the optimum level except for a few days of winter and rainy seasons. Therefore it is a problem for the farmer to bring down the raised temperature rather than to boost up. Such adverse effects of higher temperature to a certain extent be mitigated through proper designing of the rearing house and by ensuring adequate ventilation and free circulation of air inside the rearing house. The temperature reading is quite opposite in summer when compared to winter and rainy seasons. This is again a problem to cut down the raised temperature. For this all the windows should be kept open during night. Further all the doors are opened early in the morning so that cool air from outside flows into the rearing room and favours to bring down the temperature. When the day temperature increases with the day all the opened doors and windows have to be closed. Further the doors and windows are covered by wet gunny cloth to bring down the room temperature. Otherwise use of air cooler will serve the purpose. On the other hand problems of high and low temperature can be solved by air conditioning which is very costly and not affordable by many sericulturists.

Regulation of Humidity

The functions of humidity are very wide and frequent not only from season to season but also within the day itself during any season. The regulation of humidity during chawki rearing is achieved using paraffin paper. foam rubber pads. covered box rearing to get optimum levels. The humidity requirement during feeding and moulting of silkworm is different. It is necessary to lower the humidity to 60 per cent during moulting in each instar for uniform moulting. If this is neglected silkworms remain under the net. show uneven growth. become susceptible to disease. bed cleaning requires much labour. missing worms increases. Further removal of paraffin paper increases the drying effect. On the contrary high humidity is maintained during feeding which prevents leaf withering. In tropical areas the temperature is usually higher than the optimum for late age rearing. Therefore humidity should not be high as both these parameters can prove highly detrimental to late age worms. If the leaf withering is due to high temperature and low humidity it is better to give more water to the leaves by sprinkling while they are in storage. It is not advisable to raise the humidity of room by sprinkling water inside the rearing house. where the temperature is already high.

Regulation of Air

It is more important from the point of physiology of silkworm. Care should be taken to allow free circulation of air through proper ventilation to keep the toxic gases at low levels. Insecticides and disinfectants *i.e.* DDT. BHC. tobacco are harmful to the worms are more highly susceptible to these chemicals than late age worms. Thus care must be taken in handling such agricultural chemicals. Air current of 1.0 meters per second during V age rearing considerably reduces larval mortality by improving ingestion. digestion. larval weight. cocoon weight and pupation rate. Though it is less important to ventilate the room during. younger states. care should be taken to remove paraffin cover and keep the rearing bed open before each feed for free circulation of fresh air. Moreover it is one of the important factor for rearing silkworms that the temperature and humidity of rearing room can be controlled by ventilation. Therefore artificial circulation of air helps to cut down the temperature and humidity. Therefore. considering all the above facts. it is advisable to rear silkworms in the dim light during the day time and in the dark at night.

- The ecological factores chiefly temperature. humidity light and air during rearing have a significant influence on the growth of larva and ultimately on cocoon crop quality.
- The influence of these factors on rearing varies in different stages of larval growth.
- Temperature directly influences the growth of the worms. The optimums temperature for normal growth of silkworm is between 230C and 280C.
- Regulation of temperature is summer. winter and rainy days is more important. This can be achieved by air cooler. hanging wet gunny cloth to the doors and windows in summer seasons. While in winter and rainy seasons it is to heat up the room using stoves.

- The effects of humidity are direct and indirect.
- The optimum humidity ranges from 70 – 85per cent for better growth of worms.
- Young age worms require high humidity. this is possible when paraffin paper. wet foam pads and box rearing is adopted.
- High humidity generally helps to prevent leaf withering
- Silkworm require fresh air. Care should be taken to avoid the injurious gases such as Co. NH4. SO2 etc. by allowing fresh air through proper ventilation.
- Silkworms are fond of dim light of 15 to 20 lux and avoid strong and darkness.

EQUIPMENT REQUIRED FOR REARING 300 DFLS

After selection of rearing house the next phase is to consider proper rearing equipment. Sericulturist must use the space available in the room for maximum possible advantage of rearing. The rearer has to utilize all the resources available in the field for procurement/preparation of equipment. Locally available material will be cheap and reduces transport charges. Metallic equipments are to be prepared locally utilizing the skills of blacksmith and suggestion of sericulture technicians.

EQUIPMENT

The following articles constitute the rearing equipment.

1. Wooden trays
2. Bamboo trays
3. Rearing stand
4. Ant wells
5. Leaf chopping knives
6. Leaf chopping board
7. Leaf chopping mat
8. Bed cleaning nets
9. Leaf Chamber
10. Feeding stand
11. Feeding mat
12. Leaf basket
13. Chopsticks. feathers
14. Dry-wet bulb thermometer
15. Hygrometer
16. Charcoal Stove or Electric heater
17. Foot cleaning tray
18. Paraffin papers and rubber foam strips
19. Wash basin – stand

20. Foot cleaning tray
21. Sprayer
22. Nylon nets against UZI fly
23. Rough Balance
24. Litter basket
25. Humidity chart
26. Air cooler

Chemicals:

1. Formaldehyde
2. Bleaching powder
3. Paraformaldehyde
4. Benxoicacid
5. Uzicide
6. Lime Powder
7. China clay
8. R.K.O.
9. Dimilin
10. Dettol or cresol

The above listed equipment is useful for rearing. cleaning. feeding. disinfection. regulation of environmental conditions. leaf storage. pest control. brushing activities. The details description and time of using different equipment listed in this chapter have already been discussed in the earlier chapter of this book. Students are advised to review them for further understanding of this chapter.

INCUBATION OF EGGS

Silkworm eggs are of two types *i.e.* hibernating and non-hibernatings eggs. Further processing of the eggs depends upon whether they are of the diepausing or the non diapausing type. Univoltine races lay only diapausing eggs Multivoltine races lay only non-hibernating eggs while the Behaviour of the eggs of the bivoltine is intermediate. Except multivoltine. uni and bivoltine race eggs are hibernating eggs which require special treatment to make them hatch. These eggs stored till the next season or awakened from diapause artificially. The eggs stored are taken out and subjected to incubation to achieve uniform hatching on desired day. This can be achieved by exposing the eggs to certain range of environment conditions. The incubation of eggs is one of the essential parameter in silkworm rearing.

INCUBATION

The quality of eggs play a vital role in silkworm rearing. The eggs are to be incubated properly which ensures maximum hatching. Incubation is a phase in protecting the activated silkworm eggs before rearing. The environment influences the stored eggs in the development of the embryo. uniform hatching of the worms. health of the larvae. mortality. yield of cocoons. Thus the silkworm eggs are kept under proper incubation process to get high and uniform rate of hatching. on the desired date. good larval health and high cocoon quality. Acid treatment: followed by ideal incubation process. The incubation room. chamber must be clean. The required chemicals.disinfectants.equipments are to be kept available. Heating or cooling arrangements must be perfect so as to maintain uniform temperature throughout the incubation room.

The eggs cards are arranged in such a way to expose all the eggs to the temperature and room humidity. Optimum humidity in the incubation room should be 80-85 percent and temperature for non-hibernating eggs and eggs after acid treatment for immediate hatching is 240-250C right from the beginning. Over-wintered eggs taken out of cold storage are to be first kept at a temperature of 150C for three days.

Then these eggs are exposed to incubation temperature of 240-250C.This avoids sudden violent change of temperature and helps in uniform development. Eggs stored after acid treatment should be kept at 15cfor 12hours and exposed to incubation temperature:

High temperature makes the eggs hatch earlier. results in a large proportion of the eggs dying or becoming weak. They do not hatch at all. Hatched worms are lighter and remain well below the normal size resulting in forming poor silk cocoons. Too low temperature prolongs incubation process. Many eggs do not hatch and hC would be very irregular.

Conditions for Incubation

Items	No. of days in incubation											
	1	2	3	4	5	6	7	8	9	10	11	12
Temperature (0C)	25	-	-	-	-	-	-	-	-	-	-	-
Humidity (%)	85	-	-	-	-	-	-	-	-	-	-	-
Illumination	0	-	-	-	-	16 hr	-	-	-	-	0	

Humidity also plays vital role in incubation. K of optimum humidity results in poor hatching. High humidity makes the worms easily susceptible to diseases and also produce trimoulters. Light intensity during incubation hastens embryo growth. development and hatching.

It is possible to stimulate and synchronize the hatching dates of all the eggs. Silkwirms eggs may be kept under a photoperiod of 16 hours daily until 30-40per cent of the eggs reach blue egg stage: At this stage eggs are kept in dark \black boxes for more uniform hatching on the next day.

If for any unexpected reason hatching has to be delayed after incubation has started. can be done to a limited extent by cold storing the eggs at 50C on the second or third day of incubation for about a week. Hatching can be delayed at blue eggs stage by cold storing for about a week.Hatching can be delyed at blue eggs stage by cold storing forr about a week at 50C. Brushing can also be delayed by cold storing the newly hatched larvae for about three days at 7.50-100C.

HANDLING OF EGGS

The incubated eggs are handled properly to get good hatching otherwise harmful effects may result. Silkworm eggs are produced in central places. *viz.* by Government grainages and or private licensed seed producers. Readers have to purchase the eggs from egg producing entres. The eggs being live material should not be subjected to adverse environmental conditions. which will result in unhealthy development of embryos.

Hence. there is a need to transport eggs properly to ensure healthy and uniform development of the eggs. Live silkworm eggs are transported from egg producing centers: to the rearer's rearing places or chawki rearing centres (young silkworm rearing centres). The silkworm eggs are live material. The development embryo inside the ege shell need delicate handling.

The embryos need oxygen which should be made available by proper circulation of air in the container used for transportation. Non-supply of air results in suffocation which affects the healthy development of the embryo. The developing embryo need optimum temperature and humidity. If the temperature is more than the optimum. it affects the development and physiological activity of the embryo.

The eggs require about 75per cent humidity. as otherwise they dessicate:. Eggs exposed to high temperature: and humidity: during egg-laying and incubation; storage of hibernating eggs at 250C for too long period or under dry and high temperature conditions; stimulation of eggs in acid treatment; contamination of eggs by pesticides. nicotine. mosquito repellant. glue or gum causes rotten eggs before head pigmentation.

Rotten eggs are dead which perish after the colouring of the serosa and before head pigmentation with sunken shell and deep cavity. Too high temperature: during incubation; to low relative humidity:; contact with pesticides or other harmful chemicals causes dead eggs after head pigmentation. In this embryo is normally formed in the egg shell but death occurs before hatching stage. Thus keeping in view of the above dangers the eggs are to be handled carefully till hatching so as to ensure uniform and more hatching.

BLUE EGG STAGE

When the eggs are incubated: under suitable conditions reach to pin head or head pigmentation stage in 7-8 days:. This first pigmentation can be seen through the eggs shell as a blue spot and this stage is called "eye spot" stage. On the following day the whole body of the embryo turns black due to the development of body pigments and appears bluish-black through the egg shell and called as "blue egg stage". These blue egg stages hatch out in all hours:.

BLACK BOXING

Eggs after reaching blue egg stage are kept in black box/paper/ cloth and kept in dark. In this way early maturing emryos are prevented from hatching and late maturing embryos are given time to develop and catch up with the early maturing ones. Thus all the eggs reach to blue egg stage. The eggs hatch out in response to phototropic stilumus. This method favours hjatching more than 90 per cent. If hatching is not uniform and only 50-60 per cent of egs hatch on the first day. Brushing can be postphoend to next day as well. If necessary hatched worms can be separated and kept in tissue paper and stored in fridge at 100C. when the second batch hatch out the refrigerated eggs mixed and brushed together. Even blue egg stages can also be preserved at 50C for 2-3 days.

3

Osmosis and Osmotic Pressure

Although the freedom of movement of the molecules of liquids is to some extent restrained by internal cohesive forces, they also possess kinetic activity. Like gases and solutes, therefore, liquids exhibit diffusion phenomena. If, for example, water is brought into contact with another liquid such as ether with which it is only slightly miscible, a slow diffusion of water molecules into the ether will occur. Simultaneously a slow diffusion of the molecules of the ether will take place into the water. Such diffusion will continue until the two liquids are mutually saturated.

OSMOSIS

This is by far the most familiar process involving the diffusion of liquids. An understanding of the dynamics of this process, and of the significance of the physical quantity termed osmotic pressure is essential to an interpretation of the water relations of plant cells and tissues.

Let us first consider a sac-like membrane of collodion is nearly filled with a strong sucrose solution and immersed in a beaker of water. The top of the sac is tightly plugged with a rubber stopper. The collodion membrane is so prepared that it is permeable to water, but impermeable or practically so to sucrose; in other words it is differentially permeable. It is also slightly elastic.

After a short time the originally limpsac becomes rigidly distended. This is due to the movement of water through the collodion membrane into the interior of the sac. Such a diffusion of water is an as a result of the entrance of water is exerted against the inside wall of the membrane. This pressure is counterbalanced by an equal and oppositely directed pressure exerted upon the solution by the distended wall. If the membrane is virtually impermeable to the solute eventually the entire system will come to equilibrium after which there is no further increase in the volume of water inside of the membrane.

When the movement of the solvent is referred to in discussing osmosis it is always the net movement which is meant. Solvent molecules will always be moving across the membrane in both directions, but except when an equilibrium has been attained, more molecules will move per unit of time in one direction than in the other. It is the presence of this differentially permeable membrane which gives to osmosis its distinctive aspect as compared with other diffusion processes. If the membrane were permeable to the solute particles, they would diffuse in one direction, while the molecules of the solvent were diffusing in the other; this would be another example of the principle of independent diffusion.

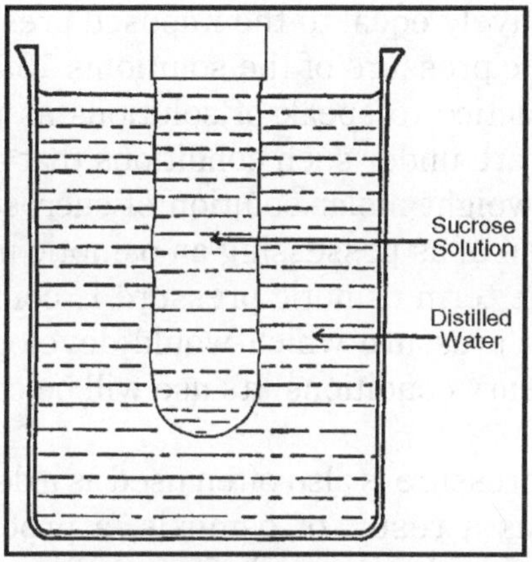

Fig. Apparatus for the Demonstration of Osmosis Through a Collodion Membrane

TERMINOLOGY

The conditions under which osmosis can occur are so varied that it is impossible to characterize the process in terms of a simple definition. The following statement will, however, apply to most cases: Osmosis is the diffusion of a liquid through a differentially permeable membrane into a solution in which the solvent is that same liquid or into another liquid with which it is miscible. In living organisms water is the only important liquid which moves from cell to cell by osmosis, hence the further discussion of this process will be in terms of water and aqueous solutions. The attainment of a satisfactory comprehension of osmosis and osmotic pressure has been delayed both by confusion in theoretical interpretations as well as by lack of a consistent terminology.

Many obscurities have resulted from the application of the term osmosis to the movement of solutes through membranes. The use of the term in this sense has not only led to many misunderstandings, hut also to positive errors and misinterpretations of data.The movement of solutes through a membrane is a diffusion process; it maybe simple diffusion, or it may be complicated by a number of factors which do not enter the phenomenon of simple diffusion. Both from the historical and theoretical standpoint it seems desirable that the term osmosis he restricted to the sense in which it is used in this discussion. Osmotic pressure may be defined as the maximum pressure which can be developed in a solution when separated from pure water by a rigid membrane permeable only to water. In order to determine the osmotic pressure of a solution it would be necessary to enclose it in a rigid membrane permeable only to water, to immerse this membrane in pure water, and to exert just enough pressure on the solution (by means of a leak-proof piston, for example) to prevent any increase in its volume due to the entrance of water.The osmotic pressure of the solution is quantitatively equal to the imposed pressure. If the latter is 15 atmos., then the osmotic pressure of the solutionis 15 atmos., etc.

It is a common practice to speak of solutions as possessing an osmotic pressure whether they are under such conditions that a pressure can develop within them or not. A weight molar solution of sucrose contained in a bottle may therefore be spoken of as possessing an osmotic pressure of 27 atmos at 25C. In other words the term osmotic pressure is commonly used to denote the potential maximum pressure which would develop in a solution were it placed under the necessary conditions. Its use will be confined to this sense in this book.

The term osmotic pressure is also often used as a designation for the actual pressures developed as a result of osmosis, a procedure which leads to confusion. Actual pressures developed during osmosis are seldom equal to the osmotic pressure as defined above. If a solution having an osmotic pressure of 12 atmos. be enclosed in a stoppered membrane permeable only to water which is in turn immersed in a solution having an osmotic pressure of 8 atmos., water will diffuse inwards until at equilibrium the actual pressure developed in the internal solution will be (disregarding any minor effect due to dilution) 4 atmos. Its osmotic pressure will still be nearly 12 atmos, however. Even if the external liquid is pure water.

The actual pressure developed in the internal solution would not be equal to its original osmotic pressure unless the membrane is completely inelastic. Inward diffusion of water results in a dilution of the solution which, as we will see later, always results in a decrease in its osmotic pressure. To avoid the inevitable confusion caused by use of the same term in a dual sense we might distinguish between the potential (*i.e.* maximum) osmotic pressure and the actual osmotic pressure of a solution. In this discussion however, we shall employ the term turgor pressure to refer to the actual pressure developed as a result of osmosis.

This term has long been employed as a name for the actual pressures developed in plant cells. Since these are largely of osmotic origin the name turgor pressure seems equally appropriate for physical systems. Actually the turgor pressure developed in a solution enclosed within a membrane may be increasing while its osmotic pressure is decreasing, as shown in the last example mentioned above. This indicates clearly that each of these terms osmotic pressure and turgor pressure refers to a different physical quantity.

In an osmometer set up as just described the maximum turgor pressure which could be developed will never exceed the osmotic pressure:

- Osmotic pressure is sometimes defined as the pressure which must he imposed on a solution to prevent any increase in its volume under the conditions specified above, but this usage is particularly unsatisfactory in the interpretation of biological phenomena. While the magnitudes of such an imposed pressure and the osmotic pressure are mathematically equal, physically they are two entirely different quantities.
- The osmotic pressure of a solution has also been variously called the osmotic potential, osmotic power, osmotic value, and osmotic concentration of a solution. Such terms are frequently encountered in the literature of plant physiology.

THE MECHANISM OF OSMOSIS

Like a gas, water or any other liquid maybe considered to possess a diffusion pressure. The only two factors which influence the diffusion pressure of a pure liquid are pressure and temperature. The imposition of pressure from an external source will raise the diffusion pressure of a liquid. For example, if a pressure of 10 atmos. be exerted upon water in a closed vessel by means of a leak-proof piston, the diffusion pressure of that water will be increased by 10 atmos.Under certain conditions negative pressures or tensions may develop in liquids. These result in a reduction in the diffusion pressure of the liquid in an amount equal to the magnitude of the tension. Although the diffusion pressure of water is also influenced by temperature no interpretation of the influence of this factor will be undertaken. In order to simplify the following discussion it will be consistently assumed that all parts of every osmotic system considered are at the same temperature. However, when a substance is dissolved in water the diffusion pressure of the water in the resulting solution is decreased as compared with that of pure water at the same temperature and pressure. This diminution in diffusion pressure is proportional, within a wide range of solution concentrations, to the number of solute particles present in a given volume (*i.e.* a given number of molecules) of the solvent.

In a solution, therefore, the diffusion pressure of the solvent may be influenced by the three factors of:

1. Temperature,
2. Pressure, and
3. The ratio of solute particles to solvent molecules.

When a solution is confined within a membrane permeable only to water and that membrane is immersed in water there will be a net movement of water through the membrane into the solution, because of the excess diffusion pressure of the water on the water side of the membrane. The passage of water through the membrane results in the development of a turgor pressure on the solution side of the membrane. Since the maximum possible pressure which can develop in the solution is equal to the excess of the diffusion pressure of pure water over the diffusion pressure of the water in the solution, the following relation holds, providing all parts of the system are at the same temperature and under the same external pressure:

Osmotic pressure Diffusion pressure of pure water Diffusion pressure of water in the solution. In other words the osmotic pressure of a solution is a measure of what may be called the diffusion pressure deficit of the water in that solution. By this term, as applied to any given solution, is meant the amount (usually expressed in atmospheres) by which the diffusion pressure of the solvent in that solution is less than the diffusion pressure of the pure solvent when the latter is at the same temperature and under atmospheric pressure.

For example, if a solution has an osmotic pressure of 10 atmos., this signifies that the diffusion pressure of the water in that solution is just 10 atmos. less than the diffusion pressure of pure water under atmospheric pressure at the same temperature; in other words its diffusion pressure deficit is 10 atmos. The osmotic pressure of a solution is a measure of the diffusion pressure deficit of the water in that solution only when the solution is not under a pressure or a tension (except, of course, atmospheric pressure) since these factors also affect the diffusion pressure of water. In plant cells, as we shall see later pressures and tensions must constantly be taken into account in evaluating the diffusion pressure deficit of the water in the cell sap.

When two aqueous solutions, both initially subjected only to atmospheric pressure, are separated by a membrane permeable only to water, diffusion of water will take place towards the solution of greater osmotic pressure. The diffusion pressure deficit of the internal solution A is 20 atmos.; that of the external solution B, 12 atmos. Water always moves in any osmotic system towards the region of its lesser diffusion pressure or in other words towards the region of its greater diffusion pressure deficit. In this particular example, disregarding dilution, the internal solution will exert a turgor pressure of 8 atmos. at equilibrium. Correspondingly the membrane will exert a wall pressure of 8 atmos. against the enclosed solution.

Since subjection to a pressure raises the diffusion pressure of water by the amount of the imposed pressure, the diffusion pressure deficit of the water in the internal solution at equilibrium will not be 20 atmos., but 12 atmos. In other words, at equilibrium, the diffusion pressure deficit of the water will be equal on both sides of the membrane.

In the preceding discussion it has been tacitly assumed that water diffuses through the membrane during osmosis in the liquid state. Many investigators consider, however, that actual diffusion of water across the membrane occurs in the form of water-vapour. Callendar who first proposed such a vapour pressure theory of osmosis, assumed the existence in differentially permeable membranes of minute capillary pores too small to be wetted by liquids, but freely permeable to molecules in the vapour state.

The essential concepts in the vapour pressure theory of osmosis can be presented in terms of two simple experiments. The presence of the sucrose molecules decreases both the diffusion pressure and the vapour pressure of the water in the solution. In fact, increase in the diffusion pressure deficit (osmotic pressure) of a solvent caused by the introduction of a solute is accompanied by a proportionate decrease in its vapour pressure.

As a result the vapour pressure just above the layer of pure water will be greater than that above the layer of the solution, and a gradient of vapour pressures will be established, decreasing in magnitude from the water to the solution. Water will therefore move from the region of high vapour pressure to the region of low vapour pressure, and the solution will steadily gain in volume at the expense of the pure water. Ultimately all of the water will be transferred to the sucrose solution. The transfer of water from one arm to the other has resulted from the difference in vapour pressures of the two liquids and is therefore directly correlated with their differences in diffusion pressures.

There will be an increase in the volume of the solution in x, and a decrease in the volume of the water in y. If we assume that the water moves through the membrane in the form of vapour there is no essential difference between this phenomenon and the one first described.In the foregoing discussion the effect of temperature on the osmotic movement of water has been disregarded.

If different parts of an osmotic system are maintained at different temperatures this will have an effect on the osmotic movement of water. However unless the difference in temperature on the two sides of the membrane is very considerable the effects of this factor onosmosis are slight, and usually can be disregarded.

The attainment of an osmotic equilibrium between two solutions, or between a solution and water does not necessarily imply an equalization of solute concentration, nor of pressure, nor of temperature. It does imply however, an equalization of diffusion pressures and an examination of the circumstances is necessary to determine the exact mechanism by which they have been brought into equilibrium in any given system.

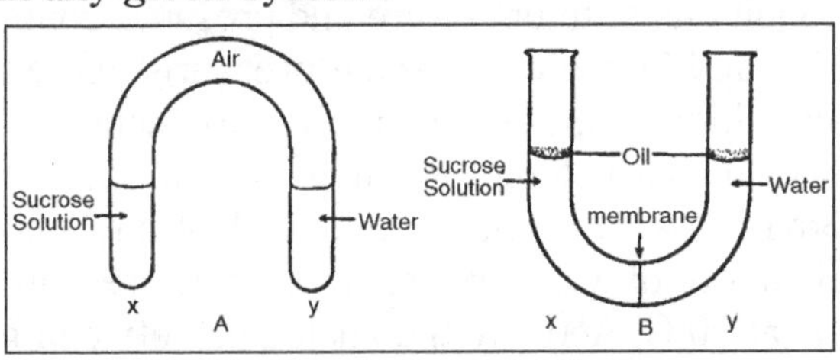

Fig. Diagrams Illustrating Discussion of the Vapour Pressure Theory of Osmosis

ANALOGIES BETWEEN OSMOTIC PRESSURE AND GAS PRESSURE

The important fact is that while the osmotic pressures of weight molar solutions of non-electrolytes approximate the same value, which is in the neighbourhood of vant Hoffs theoretical value of 22.4 atmospheres, the values for weight molar solutions of electrolytes are much higher. Vant Hoff showed that a remarkable analogy exists between the osmotic pressures of dilute solutions and the pressures exerted by gases. Vant Hoff pointed out that the osmotic pressure of a dilute solution is equal to the pressure that the solute alone would exert were it present as a gas at the same temperature in the same volume as that occupied by the solution.

A dilute solution was considered by vant Hoff to be one in which the volume of solute was small in proportion to the volume of solvent. For example, if we could imagine the complete and instantaneous removal of all of the water in a sucrose solution without any change in the spatial relations of the solute molecules, the residual sucrose molecules would be in a gaseous state. According to vant Hoff the osmotic pressure of this solution would be equal to the gas pressure which would be exerted by such a hypothetical sucrose gas. Furthermore, he showed that the osmotic pressures of dilute solutions obey laws analogous to those which describe the relations of gases to variations in temperature volume and pressure.

The analogies will now be briefly summarized:

- According to Boyles Law the pressure exerted by a gas varies inversely as its volume, temperature remaining constant. In other words the pressure of a gas varies directly with its concentration, since the smaller the volume which contains a given number of gas molecules, the greater the concentration of the gas. A similar principle holds for osmotic pressures. A molar weight of an undissociated solute dissolved in 1000 g. of water.
- According to the law of Gay-Lussac the pressure exerted by a gasaries directly with the absolute temperature, if the volume of the gas remains constant. The osmotic pressure of a solution is therefore proportional to the absolute temperature.
- According to Avogadros Hypothesis equal volumes of all gases under identical conditions of temperature and pressure contain equal numbers of molecules and hence exert equal pressures. One mol of any gas at standard conditions C.; 1 atmos. pressure) occupie 4 litres. If this gas be compressed to a volume of 1 litre, it will exert a pressure of 22.4 atmos. (Boyles Law). Similarly one mol of any undissociated solute dissolved in a litre of water exerts an osmotic pressure of approximately 22.4 atmos. at 00 C. Actually the analogies between gas pressures and osmotic pressures as discussed by vant Hoff hold with a fair degree of approximation for solutions up to concentrations of at least one molar.

This is the same pressure that it would exert were it a gas confined in the same volume under standard conditions. It follows therefore that equal osmotic pressures of solutions of undissociated solutes indicate equal concentrations of solute molecules and that Avogadros Hypothesis may be applied to osmotic pressures as well as to gas pressures. In other words the theoretical osmotic pressure of a weight molar solution of any undissociated solute is 22.4 atmos. at 00 C.

Electrolytes, as first shown by De Yries in 1884 exhibit greater osmotic pressures than their molar concentrations would lead one to expect. This fact remained unexplained until the formulation of the ionization theory by Arrhenius in 1887. In fact DeVries data were used by Arrhenius as strong evidence for his theory. Apparently each ion produced by the dissociation of an electrolyte is just as effective osmotically as an undissociated molecule. Hence all electrolytes produce higher osmotic pressures than the theoretical values for undissociated compounds; the exact magnitude of the pressures produced will depend upon their degree of dissociation.

For example, a molar solution of NaCl is about 75 per cent dissociated. This means that 75 out of every 100 molecules of NaCl present have dissociated into ions. There will therefore be 1.75 times as many particles (ions and molecules) present as in a molar solution of a non-electrolyte. The osmotic pressure of a molar solution of NaCl would therefore be theoretically 22.4 × 1.75 39.30 atmos.

The hydration of molecules or ions also influences the osmotic pressure of a solution. Water molecules which are bound to solute particles as water of hydration are no longer effective as a part of the solvent. In effect a solution containing hydrated solute particles is therefore more concentrated than its molarity would indicate. This accounts, at least in part, for the fact that the experimentally determined values of osmotic pressures are often in excess of the theoretical ones which would be expected according to the Vant Hoff analogy with the gas laws. For example, a molar solution of sucrose should theoretically have an osmotic pressure of 22.4 atmos, at o0 C.Actually its measured pressure is found to be 24.83 atmos.This discrepancy is believed to be due to the hydration of the sucrose molecules. One sucrose molecule binds six molecules of water of hydration.

Hence one mol of sucrose will bind six mols of water. One thousand grams of water is equivalent to 55.5 mols of water. Since the water of hydration is not free to act as a solvent, in effect the solution consists of 1 mol of sucrose dissolved in 49.5 mols of water. A simple calculation (22.4 ´ 55.5 49.5) yields the value 25.12 atmos. as the theoretical osmotic pressure of a weight molar sucrose solution if hydration of the sucrose molecules is taken into account. This value is in close agreement with those which have been obtained experimentally. Vant Hoffs demons-tration of the analogy between osmotic pressure and gas pressure led many subsequent investigators to consider that osmotic pressure results from the impacts of the solute molecules against the interior wall of the membrane, an assumption which vant Hoff himself was careful to avoid. One reason for the persistence of this hypothesis of osmotic pressure has been the difficulty in understanding why if the osmotic pressure is due to the solvent molecules it should be approximately equal to the pressure which would be developed if the solute molecules were present in the same volume in the form of a gas.

The solvent pressure hypothesis holds that the osmotic pressure which develops in an osmometer or similar closed system is due to the excess of the inwardly directed diffusion pressure of the solvent over the outwardly directed diffusion pressure of the solvent. Haldane has shown that this excess inwardly directed diffusion pressure of the solvent is mathematically equal to the pressure which the solute molecules inside the osmometer would exert if present as a gas in the same volume as the solution. Thus it has become possible to reconcile the concept that osmotic pressure is due to the solvent molecules with vant Hoffs classical discoveries regarding the analogy between osmotic and gas pressures.

THE MEASUREMENT OF OSMOTIC PRESSURES

The osmotic pressure of a solution, as such, can be measured only by the direct manometric method. Precise results can be obtained by this method only at the cost of an elaborate technique and infinite precautions. Hence the actual number of measurements of osmotic pressures which have been made by this direct method are very limited. As is well known there is a direct proportionality between the osmotic pressure, lowering of the vapour pressure, elevation of the boiling point, and depression of the freezing point of solutions. Hence osmotic pressures of solutions can be calculated from the determined results of one of these three physical quantities. Most frequently determinations are made of the depression of the freezing point of a solution, and its osmotic pressure calculated from this. Since the theoretical freezing point depression of a weight-molar solution of an unionized substance is 1.860 C., and the theoretical osmotic pressure of such a solution is 22.4 atmos., an equation relating freezing point depressions and osmotic pressures is easily derived. If we let A represent the freezing point depression of a solution, its osmotic pressure (O.P.) may be calculated as follows:

$$O.P.: 22.4 = A: 1.861.86O.P. = 22.4\ A22.4$$

$$A1.86O.P. = 12.04\ A4$$

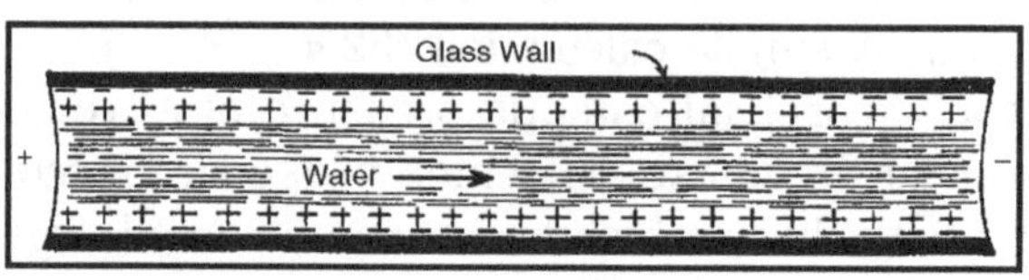

Fig. Diagram to Illustrate Electro-osmosis.

For example, a solution for which the determined freezing point depression is 0.930 would have an osmotic pressure of 11.10 atmos. (12.04 × 0.930).This equation appears to be approximately correct over a wide range of concentrations since deviations from the theoretical osmotic pressures of solutions are accompanied by almost strictly proportional deviations from their theoretical freezing point depression values. This method is often called the cryoscopic method of determining osmotic pressures. The osmotic pressures as determined by this method are as at the freezing point temperature of the solution.

ELECTRO-OSMOSIS

Under certain conditions pure water will diffuse from one side of a membrane to the other under the influence of a difference of electrical potential, a process which is called electro-osmosis. When water is confined in a capillary glass tube adsorption of OH (or HCO_3) ions imparts a negative charge to the walls of the tube. Adjoining them is a layer of H_1 ions equal in number to the adsorbed anions. In other words ions become distributed as an electrical double layer just as they do around a colloidal particle.

If a difference of electrical potential is present between the two ends of the tube, water will travel towards the negative electrode. The anions are firmly adsorbed by the walls of the tube and cannot move. The hydrogen ions, however, are free to move and migrate towards the cathode carrying the water, of which they are a part, along with them. In other words the moving A slightly more accurate form of this equation has been derived by Lewis.

The tables of Harris and Gortner based on the Lewis equation, are very useful for converting freezing point depressions into osmotic pressures. Cations in the water glide past the stationary anions which adhere to the walls of the tube. Since many membranes are essentially porous in structure,conditions similar to those just described may exist in each capillary of the membrane, and electro-osmosis may occur through such membranes in essentially the same way that it occurs through tubes of small bore. Since differences of electrical potential are frequently present in living organisms it seems entirely probable that electro-osmotic flow of water occurs in living tissues.

4

Importance of Imbibition

If a handful of the seeds of any species which do not possess seed coats impermeable to water be dropped into a vessel of water, within a few hours they will have swollen visibly. Many other materials will swell in a similar way when immersed in water. Among these are starch, cellulose, agar, gelatin and kelp stipe. Some substances will swell similarly when immersed in other liquids. All of these phenomena are examples of the process usually called imbibition. The amount of water which may enter substances in imbibition is often very great in proportion to the dry weight of the substance which swells. A piece of dried kelp stipe, for example, can absorb as much as fifteen times its own weight of water.

Mater may be imbibed as a vapour as well as in the liquid state. The swelling of doors and wood work generally during damp weather is a familiar example of this phenomenon. Plant structures, if sufficiently low in water content, also imbibe water-vapour. The water content of air-dry seeds, for example, generally fluctuates with the vapour pressure of the atmosphere, rising with an increase in vapour pressure, and vice versa.

THE DYNAMICS OF IMBIBITION

Imbibition is usually considered to be basically a diffusion process, but capillary phenomena are probably also involved. Imbibing substances are often permeated with minute submicroscopic capillaries, and it is impossible to determine how much of the liquid enters by diffusion and how much by capillary movement through invisible pores. Fundamentally, however, the cause of imbibition may be regarded as a difference in the diffusion pressure between the liquid in the external medium and the liquid in the imbibant. As long as the latter is less than the former movement of water into the imbibing substance will continue. An equilibrium will be reached, as in diffusion or osmotic phenomena, only when the diffusion pressure of the water in the two parts of the system has attained the same value.

A substance which imbibes water does not necessarily imbibe other liquids. Dry kelp stipe, for example, swells enormously when immersed in water but does not swell when immersed in ether or other organic liquids. Contrariwise, a piece of rubber does not imbibe water, but does imbibe appreciable amounts of ether and other organic compounds when in contact with themin either the liquid or the vapour state. Obviously a difference in diion pressures between the liquid in an imbibant and in its surrounding medium is not the only requisite for the occurrence of imbibition. Certain specific attractive forces between the molecules of the imbibant and the imbibed liquid must also be present. In default of such specific attractions imbibition fails to occur, even if all other necessary conditions for the process are fulfilled. While perhaps not an invariable principle it appears to be generally true that only strongly polar substances imbibe strongly polar liquids such as water, while non-polar substances such as rubber imbibe only non-polar or very weakly polar organic liquids.

Since in living organisms water is the only liquid imbibed, the further discussion will disregard other types of imbibitional phenomena. The physical relationship between the imbibed water and the imbibant is undoubtedly a complex one. The bulk of the imbibed water in any system is probably adsorbed on the molecules or micelles which constitute the structural units of the imbibing material. The possible relation of adsorbed water to the adsorbing particles in colloidal systems have already been considered.

There is good reason for believing that similar physical relations exist between the imbibed water and unit particles in an imbibant. Such water may be pictured as either actually in solution in the adsorbing particles, or as adsorbed as a shell of from one to many molecular layers in thickness on their surface. It is possible that either or both of these conditions obtain in many systems containing imbibed water, but the second of these pictures appears to be more likely.

It is also probable that a portion of the imbibed water enters and occupies minute submicroscopic capillaries which ramify between the component molecules or micelles of the imbibant. The result of imbibitionis often the production of a system which in its essential properties may be regarded as a gel. The physical status of the water in a system into which it has been imbibed and in an elastic gel is undoubtedly very similar. In fact the imbibition of a dispersion medium by a dry substance is generally regarded as a method of gel formation.

VOLUME CHANGES IN IMBIBITION

The volume of the imbibant always increases during imbibition. The final volume of the entire system (liquid+ imbibant) is always less, however, than the sum of the initial volume of the liquid and the imbibing substance. In other words a contraction in the volume of the system, liquid + imbibant, occurs during the process of imbibition. This volume shrinkage is not principally due to the occupancy of minute spaces within the substance of the imbibing material by the liquid, as might be surmised, although in some systems a part of the contraction in volume possibly may be accounted for in this way. The explanation is undoubtedly to be ascribed to the fact that the adsorbed water molecules are definitely oriented in relation to the adsorbing surfaces, and hence occupy less space than when in the free state. This is equivalent to a compression of the adsorbed water so that its density is greater than that of free water. Energy Relations of Imbibition. The process of imbibition always results in the release of heat. Liberation of heat during imbibition may be easily detected by allowing dry starch or some other material with a high imbibitional capacity to imbibe water while contained in a calorimeter, and noting the change in temperature. The adsorption of water molecules which occurs when they are imbibed results in a loss of a large part of their kinetic energy, which reappears in the system as heat energy.

This increase in the temperature of the system corresponds to an increase in the kinetic activity of all except the adsorbed molecules. The essential energy change in the process of imbibition is a loss of kinetic energy by the adsorbed molecules, and its transference to the other molecules of the system. Their increased kinetic activity is the cause of the observed rise in temperature. The greatest evolution of heat accompanies the initial stages of imbibition. This is to be expected, as a larger proportion of the first molecules of water imbibed would be firmly adsorbed than of those which enter the imbibant later.

FACTORS INFLUENCING IMBIBITION

In discussions of the factors influencing imbibition it is often necessary to distinguish between the influence of a factor upon the rate of imbibition, and upon the total amount of water which will be held in the system when an equilibrium has been attained. The principal factors which influence the entrance of water into an imbibant are as follows:

TEMPERATURE

The rate of imbibition increases with increase in temperature. Equilibrium will therefore be attained more slowly at lower temperatures, but the amount of water held in an imbibant at equilibrium will be slightly greater at lower than at higher temperatures. This is in accord with the principles of the influence of temperature upon adsorption processes generally.

OSMOTIC EFFECTS

Water moves by imbibition into a substance only when its diffusion pressure is in excess of the diffusion pressure of the water in the imbibant. The introduction of a solute into water invariably has the effect, as already discussed, of reducing the diffusion pressure of the water in that solution as compared with pure water. The osmotic pressure of a solution is a measure of its diffusion pressure deficit. Hence the magnitude of the osmotic pressure of the surrounding solution would be expected to influence both the rate of imbibition and the equilibrium water content of an imbibant.

One of the most comprehensive studies dealing with osmotic effects in relation to imbibition has been conducted by Shull who used the seeds of the cocklebur (Xantlziurn pennsylvanicum) as the imbibing material. Samples of the seeds of this species were allowed to come to equilibrium with solutions of sodium chloride and lithium chloride of different osmotic pressures, after which the amount of water which had been imbibed was determined. The coats of these seeds are permeable to water, but virtually impermeable to these solutes. Were it not for this latter fact complicating effects due to the specific influences of the ions employed, as described in the next section, might invalidate the results of experiments of this type. The general principle shown by the results of these experiments is that with increase in the osmotic pressure of the solution in which the seeds are immersed, the amount of water held by imbibition per unit of dry weight at the equilibrium point decreases.

Since at equilibrium the diffusion pressures of the water in the imbibing substance and in the surrounding liquid must be equal the basis for this osmotic effect upon imbibition is evident. If the diffusion pressure in the solution is relatively low (high osmotic pressure) the diffusion of less water into the imbibing substance is necessary to raise the diffusion pressure of the water in it to a value equal to the diffusion pressure of the water in the circum ambient liquid, than if the diffusion pressure in the surrounding liquid is high.

The relation between the osmotic pressure and the amount of water imbibed is not, however, a strictly proportional one. At the lower end of the range of concentrations employed, an increase of a few atmospheres in osmotic pressure causes a decrease in the amount of water imbibed, in terms of the air-dry weight of the seeds, of nearly15 per cent. Near the upper end of this range of concentrations an increase in osmotic pressure of several hundred atmospheres is required to produce an equivalent change in the volume of water imbibed. This is further evidence that the first increments of water passing into any imbibant are held by tremendously greater forces than those which are imbibed subsequently.

IONIC EFFECTS

The amount of water which will be imbibed from a given solution may also be influenced by the species of ions in that solution. If agar or kelp stipe, for example, is allowed to come to equilibrium with equimolar solutions of lithium, sodium, ammonium, and potassium chlorides swelling will not be equal in the different solutions. The greatest imbibition will occur in lithium chloride solution, next greatest in sodium chloride, next greatest in ammonium chloride, and least in the potassium chloride solution. In all of the solutions swelling will be less than in pure water.

At first sight it would seem that this might be ascribed to an osmotic effect.The potassium chloride solution might be assumed to have a higher osmotic pressure than the lithium chloride solution, due to differences in dissociation or hydration, and hence to exert more of a retarding effect upon imbibition than the other solutions. Actually the opposite condition holds, hence this influence of electrolytes upon imbibition cannot be explained in terms of an osmotic effect. Apparently, since all of these electrolytes give rise to the same anion the cause of this phenomenon must be looked for in terms of specific effects of the cations upon swelling.

Many other processes and physical properties are affected by this and similar ionic sequences. Such ionic series as these are known as lyotropic series, and such effects of ions as are found to be arranged in sequences of this sort are called ljo tropic effects.

Sometimes the order of the effect of a lyotropic series is from left to right and sometimes the reverse depending upon the property affected. As arranged above the series of univalent cations represents (reading from left to right) the order of their increasing effectiveness in checking the imbibition of water by kelp stipe. The same is true of the series of bivalent cations.The anions in the third series do not differ very greatly in their effects upon the swelling of kelp stipe, but are known to influence many other processes the magnitude of their effects being in the order given.

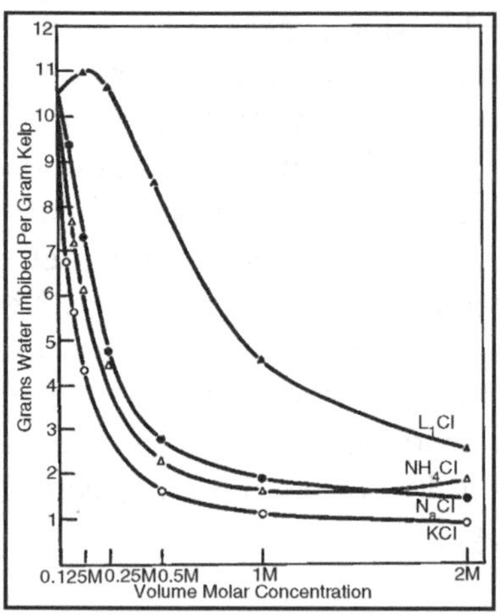

Fig. Influence of Cations on Imbibition of Water by Kelp Stipe.

These so-called lyotropic effects are apparently due largely to differences in the degree of hydration of various kinds of ions although other factors are probably also involved. Their lyotropic effect on the swelling of kelp stipe, agar, and other similar materials which bear a negative charge with respect to water probably may be explained as follows: The negative charge on the kelp stipe is apparently due to adsorption of hydroxyl ions, leaving hydrogen ions in the outer zone of the electrical double layer. When immersed in a solution of an electrolyte not only is water imbibed, but an exchange of cations occurs, some of the cations in the electrolyte replacing hydrogen ions in the kelp stipe. The displaced hydrogen ions of the double layer pair off with the residual anions of the electrolyte. The water molecules associated with the cations which enter the double layer will contribute to the total hydration of the kelp. The different degrees of swelling caused by the different cations is probably due partly to the different numbers of water molecules associated with individual cations of various species, and partly to the replacement of a larger proportion of hydrogen ions in the double layer by some species of cations than others.

In addition to this lyotropic effect upon swelling, ions often exert an electrostatic effect as well. Solutions of chlorides producing bi- and trivalentcations ($CaCl_2$, $AlCl_3$, etc.) result in a great reduction in the swelling of kelp stipe even when the concentration is so dilute that their osmotic effects are negligible. Exchange of the polyvalent cations for the H + ions in the double layer apparently results in a type of flocculating effect upon the negatively charged particles of the kelp.This brings the particles closer together and results in less swelling. Such effects are also exerted when polyvalent cations are associated with other univalent anions, but when paired with polyvalent anions the decreasing effect of salts producing polyvalent cations upon the swelling of negatively charged imbibants is less marked. There is a close similarity between this phenomenon and the effect of electrolytes in flocculating negatively charged sols.

EFFECTS OF H-ION CONCENTRATION

Hydrogen ions in any appreciable concentration affect the swelling of a negatively charged imbibant adversely while hydroxyl ions in any appreciable concentration cause greater swelling of such systems than occurs in pure water. Exactly the reverse effects are exerted by hydrogen and hydroxyl ions upon the swelling of positively charged imbibants. The swelling of amphoteric gel-forming substances such as gelatin and certain other proteins is influenced in an even more complex manner by the pH of the swelling medium. Minimum swelling occurs at the isoelectric point, and with increase or decrease of pH from this point increased swelling occurs. At relatively low or relatively high pH values a decreasing effect upon swelling is again evident.

IMBIBITION PRESSURE

Pressures, sometimes of an enormous magnitude develop during the swelling of an imbibing substance. Such pressures only become evident if the imbibant is confined in some way during the process of imbibition. One common method of demonstrating the development of pressures during imbibition is as follows:

A glass funnel, lined with filter paper is partly filled with moist plaster of Paris paste. The surface of the plaster of Paris is then strewn with a number of pea seeds, after which more paste is added until the funnel is full. In a few minutes the moist matrix containing the seeds will have set. The resulting solid cone is removed from the funnel and its base immersed in a dish of water. Water moves upward through the porous gypsum cone by capillarity and permits a continued imbibition by the pea seeds. Within a few hours the pressure developed by the swelling pea seeds is sufficiently great to rupture the gypsum block.The magnitude of the pressure developed during imbibition may be measured quantitatively in two different ways.

It may be counterbalanced by a mechanical pressure, the magnitude of which becomes a measure of the imbibition pressure developed. Such a method was used by Reinke as long ago as 1879. He stacked disks of dried fronds of Laminaria (a sea weed) in a hollow metal cylinder, and inserted above the disks a metal piston bearing a platform at its upper end. Water was then brought into contact with the dry disks. By placing upon the platform weights of sufficient mass to just prevent swelling of the kelp, the magnitude of the imbibition pressure was determined.

A second method of measuring this quantity relies upon counterbalancing the imbibition pressure of a substance by means of a solution of high osmotic pressure. The pressure developed by imbibition may be regarded as due to the diffusion pressure of the entering water. If the imbibing substance be immersed in a solution in which the diffusion pressure deficit (osmotic pressure) is sufficiently great to prevent any movement of water into that substance the osmotic pressure of that solution is a measure of the imbibition pressure which would be developed in that imbibant. A solution with an osmotic pressure of nearly a thousand atmospheres was required to prevent any imbibition of water by cocklebui seeds. As determined by either of the two methods just described the imbibition pressure of an imbibant is analogous to the osmotic pressure of a solution, *i.e.* it is a measure of the maximum imbibition pressure which the imbibant can develop.

5

Effects and Impact of Permeability

According to the principles of diffusion, the molecules and ions of all substances in solution, and even colloidal micelles, tend to attain an equal distribution in terms of concentration through all parts of a system. The same is true of gases, and the osmotic movement of water, as we have seen earlier is a response to the same general principle. In living organisms, however, such an equality of distribution of the particles of a solute among the various cells, or between cells and a liquid environment, is seldom attained. There is a marked heterogeneity in the distribution of solutes among the different cells in tissues, in entire organisms, and even among the different parts of the same cell.

The diffusion of substances in solution into or out of plant cells is limited and modified in a number of ways. The diversity of the distribution of solutes within the plant body is a result of conditions which modify or check their free diffusion. The permeability of the cell, or some part thereof, is one of the most important factors influencing the rate with which molecules and ions diffuse into or out of plant cells. Plant cells and parts of plant cells such as nuclei and plastids are bounded by membranes which are notable for their differences in permeability to different substances.

The permeability of a membrane to a given substance is measured by the rate at which that substance will passthrough that membrane under a given driving force. In the simplest cases, this driving force depends upon the steepness of the concentration or diffusion pressure gradient of the diffusing substance across the membrane.The concept of permeability is inseparable from the idea of a membrane.Permeability, it should be clearly understood, is a property of the membrane and not of the substance which diffuses through it.

Common examples of non-living membranes, some of which have already been described in previous chapters, are thin sheets or layers of rubber, collodion, parchment paper, cellophane, gelatin, and copper ferrocyanide. Liquids may also serve as membranes under certain conditions, as an example described later in this chapter shows. Most biologically important membranes are of the type characterized as differ-entially permeable. Different substances diffuse through such membranes at different rates. They may be, and often are, impermeable or virtually so to some substances, while others pass through them readily. Such membranes are sometimes termed selectively permeable or semi-permeable, but these two terms are not as truly descriptive of their properties as the designation differentially permeable.

THE MEMBRANES OF PLANT CELLS

With very few exceptions the protoplasm of plant cells is enclosed by a definite, more or less rigid cell wall. This cell wall fulfills all the criteria for being considered a membrane. Through it must pass all substances moving either into or out of a cell. The cell wall membrane is sometimes called the non-living membrane of a cell but the assumption that all cell walls are entirely non-living is scarcely a validone. The plasmod-esms, at least, must be considered in any problem involving the permeability of the cell walls of living plants. Lining the cell wall of all mature plant cells is the layer of cytoplasm. The cytoplasm, or parts thereof, comprises the second membrane through which substances entering or leaving the vacuole of a cell must penetrate.

This layer is sometimes called the living membrane of the cell. Both theoretical considerations and experimentally observed facts support the view that the two limiting layers of the cytoplasm possess different physicochemical properties from the interior of the cytoplasm and actually may be regarded as distinct membranes. The cytoplasmic membrane adjacent to the cell wall is called the plasmalemma; that enclosing the vacuole the tonoplast or vacuolar membrane.

Since the boundaries between the cytoplasm and the cell wall and the cytoplasm and the vacuole constitute interfaces, protoplasmic ingredients which lower interfacial tension probably become more concentrated in these layers than in the interior of the cytoplasm. For this reason alone the postulation that the cytoplasmic membranes possess different properties from the interior cytoplasm seems justified. Results from several types of experimental investigations support this hypothesis. The immiscibility of protoplasm with water, already discussed, appears to be at least partly due to the fact that protoplasm is coated with a layer of material which is insoluble in water.

By means of micropipettes certain non-toxic dyes can be injected into the body of the cytoplasm through which they will spread rapidly.They do not, however, pass through either the plasmalemma or the tonoplast. Neither will they pass into the interior of the cytoplasm from a solution bathing the cell, nor from the vacuole if the latter is injected with the dye.

This is direct evidence that the constitution of the plasmalemma and the tonoplast is different from that of the interior cytoplasm. The presence of cytoplasmic membranes can also be demonstrated by the manipulation of living protoplasm with fine glass needles. It has been found possible to withdraw a plasmolyzed protoplast from its cell wall and to strip off the protoplasm from the vacuole leaving the latter as a free floating sac full of cell sap enclosed by a delicate membrane-the tonoplast. Most evidence indicates that the tonoplast and probably also the plasmalemma are thin liquid films which are immiscible with water. A complete picture of the permeability of a cell to different. Removal of the outer cytoplasm compounds can be drawn only in from the tonoplast and vacuole of a plasmo terms of the permeability of both lyzed cell.

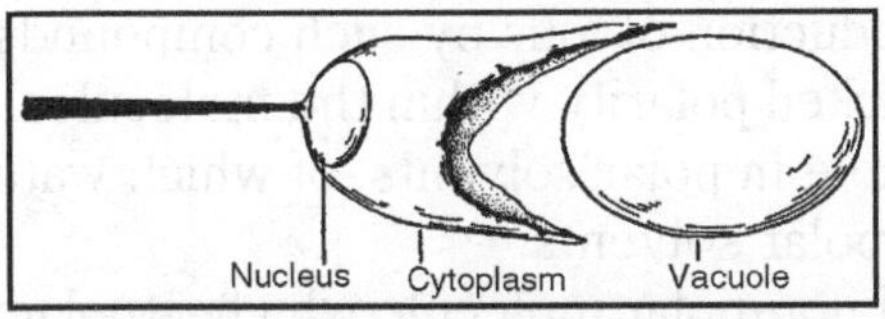

Fig. Removal of the Outer Cytoplasm from the Tonoplast and Vacuole of a Plasmolyzed Cell

In entering a cell a substance must first pass through the various layers of the cell wall, thence in turn through the plasmalemma, the interior cytoplasm, and the tonoplast before reaching the vacuole. Although some investigators prefer to refer all permeability phenomena of plant cells to the cytoplasm as a whole, the evidence that the plasmalemma and the tonoplast exist as distinct membranes with the property of differential permeability must be regarded as very substantial.

It is also customary to speak of certain plant structures composed of one or more layers of cells as membranes. Examples are the seed coats of many seeds and fruits (especially grains) epidermal layers such as those peeled from the scales of an onion bulb, the skin of potatoes, etc. Such membranes may be termed for convenience multicellular membranes. They maybe either living, as for example onion epidermis, or non-living, as for example many seed coats.

THE PERMEABILITY OF THE CYTOPLASMIC MEMBRANES

The cytoplasmic membranes are differentially permeable membranes par excellence. The term cytoplasmic membranes will be employed in a loose sense to refer to the entire cytoplasmic system of membranes: plasmalemma plus interior cytoplasm plus tonoplast. Actually many substances entering plant cells are intercepted and utilized in one way or another in the cytoplasm, and never diffuse through the tonoplast. Similarly compounds synthesized in the cytoplasm may pass out of the cell without crossing the tonoplast. Hence the plasmalemma is often regarded as the most important unit in the cytoplasmic system of membranes. A pertinent concept as a background for a discussion of the permeability of the cytoplasmic membranes is the distinction between polar and Non- polar compounds and groups. Water, electrolytes, carbohydrates, and amino acids are typical strongly polar compounds. This polarity is fundamentally an electrical phenomeNon- and is due to the grouping of the atoms in a molecule, as well as to the actual structure of its constituent atoms in terms of the spatial arrangement of protons and electrons. A polar molecule can be likened to a tiny magnet since it possesses certain regions which are electro-positive, while other regions are electro-negative.

This property of polarity is more strongly marked in some species of molecules than in others. Electrolytes possess the most strongly polarized of all molecules; the production of ions by such compounds may be considered a manifestation of a marked polarity within the molecules. Polar compounds are in general, more soluble in polar solvents, of which water is the most familiar example, thanin non-polar solvents.

The most typical non-polar molecule which is polar at one end; non-polar at the compounds are the hydrocarbons of both the other, chain and ring types, such as methane (CH4) and benzene (C_6H). Non-polar compounds are much more soluble in non-polar solvents such as benzene, chloroform, and ether than in polar solvents such as water.

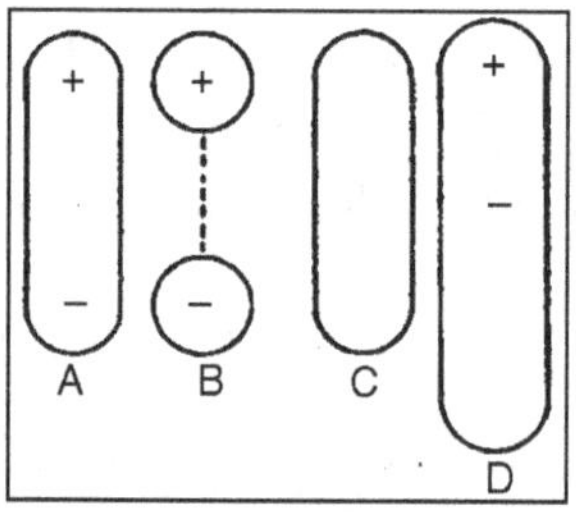

Fig. Diagrams Illustrating.

(A) Polar molecule,
(B) Ionized polar molecule,
(C) Non-polar molecule,
(D) Molecule which is polar at one end; non-polar at the other.

The concept of polarity applies not only to molecules as a whole, but to certain atomic groupings within molecules. Such groups as CH_3, C_2H_5, etc., are non-polar, while COOH, OH, NH_2, CHO, CN are the more familiar of the polar groupings. The molecules of many types of compounds, as for example the organic acids, contain both polar and non-polar groups. One end of an acetic acid (CH_3COOH) molecule, for example, is decidedly polar, while the other is essentially non-polar. In general, the greater the number of non-polar groups which a molecule contains in proportion to the number of polar groups present the more predominant will be the non-polar properties of the molecule and vice versa.

One of the few important generalizations which may be made regarding the permeability of the cytoplasmic membranes is that they are usually much more permeable to predominantly non-polar than to predominantly polar molecules.

PERMEABILITY OF THE CYTOPLASMIC MEMBRANES

Electrolytes are all strongly polar compounds, hence it is not surprising to find that the cytoplasmic membranes of many plant cells are not readily permeable to them. However, as described later in this chapter electrolytes often enter certain kinds of plant cells readily. The actual mechanics of their penetration through the cytoplasmic membranes is a more complex process than simple diffusion. Some authorities contend that electrolytes penetrate as ions; others that they penetrate only as molecules. For the purpose of this discussion the first of these two viewpoints will be adopted.

Univalent cations such as Na+ and K1 usually penetrate through the cytoplasmic membranes more readily than bi- and trivalent cations such as Ca+ + and Al+ + +. The same general relationship also seems to hold for anions. Among non-electrolytes in general the less polar the molecules, the more readily they can penetrate through the cytoplasmic membranes. Hydrocarbons such as benzene, being the least polar of all compounds, penetrate into cells very rapidly, although in only very small quantities, because of their low solubility in water. The cytoplasmic membranes are usually quite permeable to the gases carbon dioxide and oxygen when in the dissolved state. This is in accord with the general principle since the molecules of these gases are essentially non polar. When carbon dioxide dissolves in water a large proportion (more than 50 per cent) of the molecules present react with the water, forming carbonic acid (H_2C0_3).

This is a complicating factor which must be considered in any evaluation of processes involving the diffusion of dissolved carbon dioxide. Most organic compounds containing only one polar group such as OHCOOH, or NH_2 diffuse across the cytoplasmic membranes less rapidly than those containing no polar groups, but more rapidly than compounds containing two polar groups. Further increase in the number of polar groups present in a molecule still further decreases the penetrability of the molecule. Glycerol (C_3H_5(OH) for example, an alcohol containing three polar OH groups, penetrates relatively slowly, while monohydric alcohols such as methyl (CH_3OH) and ethyl (C_2H_5OH) penetrate through the cytoplasmic membranes very rapidly. Most studies indicate that the cytoplasmic membranes of most kinds ofcells are usually relatively impermeable to sugars. Impermeability to sucrose is especially marked. Since molecules of sugars contain a large number of polar groups this is in accord with the general conception of the relation of the polarity of molecules to the permeability of the cytoplasmic membranes to them. However, as in the case of electrolytes, there are some kinds of cells which sugars enter relatively rapidly. Amino acids another extremely important group ofcompounds physiologically, also penetrate through the cytoplasmic membranes of most cells relatively slowly. Since each amino acid molecule contains oneor more

NH_2 groups and one or more COGH groups the behaviour of these compounds is in accord with the general principle. Water molecules, although decidedly polar, are an exception to the general rule, usually penetrating through the cytoplasmic membranes with relative ease. This is probably possible only because of the relatively small size of the water molecule. That cytoplasmic membranes are often, at least, quite permeable to water is shown by the frequent rapid loss or gain of water by plant cells.

Another important difference between permeability to polar compounds and to non-polar compounds is that ease of penetration of the former shows a close correlation with particle size, while this is not true of the latter. As among different ions of the same electrostatic charge, for example, the smaller ions (due account being taken of hydration) seem to penetrate more rapidly. No such correlation is evident between particle size and permeability among molecules of a predominantly non-polar structure.

In fact, exactly the reverse relation sometimes holds. For example, in the saturated series of fatty acids the larger the number of non-polar groups present, *i.e.* the greater the size of the molecule, at least up to a certain limit, the greater the permeability of the cytoplasmic membranes to the acid. According to Loeb the order of penetrability of the saturated fatty acids into sea urchin eggs is asfollows, the most rapidly penetrating acid being given first: nonylic ($CH_3(CH_2)_7GOGH$), caprylic ($GH[(GH_2)_6$ GOGH), caproic ($GH_3(CH_2)_4$ GOGH), butyric (GHy $(CH_2)_2GOGH$), proprionic (CH_3GH_2GOGH), acetic (GH_3 GOGH) and formic (H. GOGH).

Permeability of the cytoplasmic membranes to polar compounds differs from permeability to non-polar compounds in one other respect. Permeability to the former is markedly influenced by external factors and by internal changes in the membrane. The same membrane apparently may vary all the way from a condition of marked permeability to certain polar solutes to a condition of almost complete impermeability to the same solutes. Permeability to non-polar solutes, on the other hand, is not only relatively high, but much more nearly constant, and much less influenced by environmental factors.

FACTORS AFFECTING THE PERMEABILITY OF THE CYTOPLASMIC MEMBRANES

A number of factors of external origin may influence the permeability of the cytoplasmic membranes. Only the more important of these will be considered. As the preceding discussion has indicated environmental factors influence the permeability of the cytoplasmic membranes to polar compounds more markedly than to non-polar compounds.

TEMPERATURE

The apparent permeability of membranes to many substances increases with temperature. In part this effect may be due to increased kinetic activity of the molecules which pass through the membrane rather than to any effect of temperature on the membrane itself. A given substance will penetrate through a membrane more and more rapidly as the temperature is raised, until the membrane itself is destroyed, or profoundly modified by the temperature attained. In living membranes the upper limit of temperature which results in the destruction of the membrane is the thermal death point of the protoplasm which in most active cells usually lies between 50 and 600 C.

At this point a sudden, irreversible increase in permeability occurs. This is due to death of the protoplasm and consequent destruction of the property of differential permeability in the cytoplasmic membranes. This phenomenon may be observed readily by heating a small piece of red beet tissue in a test tube of water. As soon as the thermal death point of the protoplasm is exceeded, the red pigment dissolved in the cell sap (an anthocyanin), to which the living cytoplasmic membranes are impermeable, rapidly diffuses out into the water.

Temperatures low enough to induce ice formation in plant tissues also result in an increase in the permeability of the cells of many (but by no means all) tissues, due to the resultant destruction of the protoplasm. This effect can be conveniently illustrated by freezing a smallblock of red beet tissue and immersing it in water as soon as it has thawed. The almost immediate outward diffusion of anthocyanins which ensues is evidence of the destruction of the cytoplasmic membranes.

ELECTROLYTES

According to Osterhout investigations made on the tissue of Larninaria (a kelp) by the conductivity method 1 indicate that the cations Li^+, Na^+, K^+, NH^{4+}, Cs1, and Rb^+ all increase permeability. This seems to be true regardless of the anion with which the cations are paired. The cations are arranged from left to right in their probable order of effectiveness in increasing permeability. This, it will be noted, is approximately the order of the lyotropic series.The initial effect of bivalent and trivalent cations such as Ca^{++}, Ba^{++},Sr^{++}, Mg^{+1}, Fe^{+1}, La^{+11}, Fe^{11+} and Al^{11+} as found by this method In this method permeability is determined by measuring the electrical conductivity of cylinders or stacks of disks of plant tissue. The greater the electrical conductivity, the greater the permeability. Since the conductivity of the tissue is due to the movement of ions through it, this method measures permeability to electrolytes only.

This primary effect is usually followed in time by a secondary increasing effect upon permeability. These results were found when these cations were associated with univalent anions. When the anions are of higher valence the initial decreasing effect upon permeability may be diminished or entirely offset.

Similar effects of univalent and bivalent cations on the permeability of the cytoplasm of the epidermal cells of onion bulb scales to water are recorded by de Haan.Raber (1920), using the same method and type of material as Osterhout studied the effects upon permeability of a number of salts composed of the same cation (Na^+) but different anions. All of these salts were found to result in an increase in permeability. When supplied in equimolar concentrations the order of the increasing effect of the anions upon permeability was: citrate] $P0_4$] tartrate] SO_4 and acetate] Cl] NO_3] Br] I]CNS.

The influence of the four polyvalent anions in increasing permeability was much more marked than the univalent anions. Both of these series follow closely the lyotropic order as usually given for anions. The effect of any electrolyte upon permeability can be analysed only in terms of the anions and cations into which it dissociates. The results of conductivity experiments seem to indicate that the following general principles hold: If the valency of the anion exceeds that of the cation the salt results in an increase in permeability. If the converse is true the initial effect of the salt is a decrease in permeability; this is followed by a secondary increasing effect. When cation and anion are of the same valence the effect will depend upon the relative size of the ions and other factors. When both ions are univalent, however, the usual effect is an increase in permeability.

The interpretation of these effects of electrolytes upon permeability is not entirely clear, but if we assume the cytoplasmic membrane to be composed of negatively charged micelles, an explanation similar to that given in the preceding chapter for the influence of ions upon imbibition would seem to fit the facts closely.Not all studies of the effects of ions upon permeability have yielded results in accord with those just discussed. For example Brooks found that sodium, potassium, magnesium, and calcium chlorides in concentrations of 0.0125 to 0.05 molar all check the penetration of the dye dahlia into living cells of Nitella. Similarly 115 in found that dilute solutions of the above salts as well as several others all result in a decrease in the permeability of the cytoplasmic membranes of the cells of various species to sugars. The effects of electrolytes upon permeability of the cytoplasmic membranes probably varies not only with their concentration but also with different types ofcells, and with the same cell under different conditions. The effects of a mixture of two entirely different salts upon permeability should also be considered. This is illustrated in an experiment described by Bayliss. A slice of freshly cut red beet was immersed in a solution of 0.31M sodium chloride. It was observed that the red pigment gradually diffused out into the solution. Since when similar slices were immersed in water no such outward diffusion of red pigment occurred this showed the effect of sodium chloride in increasing permeability. If, however, enough calcium chloride was added to a portion of the sodium chloride solution before the slice of beet was placed in it to bring the concentration of the former salt to 0.17 M no such outward diffusion of red pigment could be detected. Evidently the presence of Ca^{++} ions in some way offset the usual effect of Na^+ ions upon permeability. Such an effect of one ion upon another is known as antagonism.

NON ELECTROLYTES

Certain non-electrolytes are known to have effects upon permeability. Among these are the anaesthetics. Exposure of the cytoplasmic membranes to extremely low concentrations of ether or chloroform has no apparent effect upon their permeability; exposure to slightly higher concentrations decreases their permeability. With exposure to still higher concentrations this decreasing effect on permeability is only temporary, and is followed by a rapid increase in permeability which may continue until the value characteristic for the death point of that tissue is reached.

This secondary effect of anaesthetics is due to the toxic effect of high concentrations upon the protoplasm resulting in its death after a relatively short period of time. The initial decreasing effect of anaesthetics upon permeability is reversible, but the secondary increasing effect is usually non-reversible.

HYDRATION

Changes in the osmotic or imbibitional relations of the water in other parts of the cell may induce changes in the hydration of the cytoplasmic and cell wall membranes. It is inferred that variations in the permeability of the cytoplasmic membranes occur with shifts in their water content. Such changes in permeability are often presumed to be related to changes induced in the viscosity of the cytoplasm due to variations in its degree of hydration. Extreme desiccation of most cells results in a coagulation of the protoplasm, and a consequent complete loss of the property of differential permeability by the cytoplasmic membranes. In some cells, however, particularly those of drought resistant plants, the cytoplasm may regain its normal condition of differential permeability when water again becomes available after a period of practically complete desiccation.

STIMULATION

This ignorance-concealing term is applied to a number of phenomena which result in an increase in permeability, the exact nature of which are so incompletely understood that they cannot be described in any more fundamental terms. For example, it has been found that the permeability of the egg cells of certain species of invertebrate animals increases when fertilization occurs. Likewise injury or wounding of cells or tissues, whether due to mechanical or other causes, also usually results in an increase in permeability.

THE PERMEABILITY OF THE CELL WALL MEMBRANES

The permeability of the cell wall membranes is greatly influenced by their chemical composition and physical organization. Cell walls composed of cellulose and pectic compounds, such as those in most parenchymatous tissues, are usually quite permeable to water, and, if wet, to many dissolved substances. Water and solutes probably pass through such walls largely if not entirely through the hydrophilic intermicellar material. The plasmodesms also undoubtedly facilitate the movement of water and solutes through the walls of many living plant cells.

That the differentially permeable membrane of such plant cells lies in the cytoplasm is visually shown when the protoplasm of any cell which contains anthocyanins dissolved in the cell sap is destroyed by heating, freezing, or toxic agents. As long as such a cell is alive and uninjured the red pigment does not diffuse out of the vacuole, even if the cell is immersed in water.

Immediately upon destruction of the protoplasm, however, outward diffusion of anthocyanins and other solutes occurs through the intact cell wall, indicating that the latter does not appreciably impede their egress from the cells. Lignified cell walls are also quite freely permeable to water and, when wet to solutes as well. Lignification apparently results in little if any decrease in the permeability of the intermicellar material.

Cell walls in which cutin or suberin are present in appreciable quantities on the other hand, are relatively impermeable both to water and solutes. While it appears to be generally true that cell wall membranes which are readily permeable to water are also readily permeable to solutes dissolved in water, a number of differentially permeable cell wall membranes have been discovered and it is probable that differential permeability is more often a property of the cell walls than is generally realised. The coats of a number of different kinds of seeds and fruits have been found to act as differentially permeable membranes.

These coats may be several cell layers in thickness but the cells of which they are composed are often dead. Even in those in which the cells are living destruction of the protoplasm by one treatment or another does not destroy the property of differential permeability. Clearly therefore, this property is inherent in the cell walls. Many such membranes are relatively permeable to water but show different degrees of permeability to solutes, ranging all the way from practical impermeability to some to a marked permeability to others.

Among the species whose seed or fruit coats have been found to show this property are many of the grains, peach, apple, bean, horse-chestnut, sugar-beet, sunflower and cocklebur. For example, Shull found that the seed coat of the cocklebur was impermeable to ether acetone, chloroform, ethyl alcohol, glycerol, sugars, NaCl, CuSO4, HCl, and tartaric acid among others, but was permeable to a greater or lesser degree to KCl, KOH, NaOH, H_2S0_4, HNO_3, $AgNO_3$, $NaNO_3$, acetic, citric, and lactic acids, as well as some others.

RELATION BETWEEN PERMEABILITY AND THE RATE OF ABSORPTION

There is not necessarily any correlation between the permeability of the cell membranes to a given compound and the rate at which that substance will enter the cell from other cells or a liquid medium. The permeability of the cell membranes does not control the rate at which compounds pass either into or out of a cell but is merely one factor which influences their rate of movement through the cell wall and the cytoplasm. For example, the membranes of a cell may be readily permeable to a given substance which nevertheless may enter very slowly because the vacuolar concentration of that substance is already relatively high. Another substance, to which the membranes are relatively impermeable, may move into the cell continuously because it is being utilized in some metabolic process within the cell.

Recent work on the absorption of ions by plant cells indicates that the results of many studies upon permeability to electrolytes are somewhat misleading. In actively metabolizing cells ions often accumulate until their vacuolar concentration is much greater than in the outside medium. The penetration of electrolytes into such cells often occurs at a relatively rapid rate, in spite of much evidence which indicates that the cytoplasmic membranes are not very permeable to them. Similarly there is evidence that sugars sometimes move into at least certain kinds of cells at a relatively rapid rate. A probable explanation of this apparent discrepancy is that most studies of permeability have been made upon cells which do not possess the capacity of absorbing electrolytes or other polar compounds at a rapid rate. This capacity seems to be restricted largely, if not entirely, to meristematic or otherwise rapidly metabolizing plant cells.

THEORIES OF MEMBRANE PERMEABILITY

A number of hypotheses have been proposed as an explanation of the mechanism of differential permeability in both inorganic and living membranes. The discussion will be limited largely to the bearing of these theories upon the permeability of the cytoplasmic membranes. The two theories which have been most favourably regarded in all discussions of this topic are the sieve theory and the solubility theory. A number of variants of each of these theories has been proposed.

SIEVE THEORY

Probably the simplest picture which can be conceived of the operation of a membrane is that it works as a molecular sieve. The pores in the membrane may be imagined to be of such dimensions that micelles molecules or ions below a certain size can diffuse through them while those exceeding this size limit cannot pass through. Certain types of inorganic membranes such as collodion, parchment paper, and copper ferrocyanide undoubtedly owe their property of differential permeability largely to such a sieve-like action. Although a sieve theory of the permeability of the cytoplasmic memb rane sis strongly advocated by some authorities, notably Bayliss there are some very grave objections to the acceptance of any such hypothesis as a complete explanation of this phenomenon. It is difficult to conceive of a sieve like structure of the cytoplasmic membranes, except insofar as this is imparted by the normal intermolecular distances which are of course, very minute. A more serious difficulty, however, is found in the fact that the permeability of the cytoplasmic membranes to certain classes of compounds actually increases with increase in the size of the molecule, at least up to a certain point. This is true in certain homologous series of organic compounds already cited.

SOLUBILITY THEORIES

A second generally advocated theory is that membranes are permeable to substances which will dissolve in them and impermeable to others. The operation of an inorganic membrane of this type can be readily demonstrated by a simple experiment. A test tube is about half filled with chloroform on top of which is introduced a very thin layer of water, and then nearly filled with ether. The water may be coloured faintly with a water soluble dye such as methylene blue in order to aid in distinguishing it from the other liquids. After stoppering, the tube is clamped in an upright position and the position of the water meniscus is marked accurately.

A second tube is prepared in exactly the same way, except that xylene is substituted for ether as the upper layer of liquid. If examined after a period of several days, it will be found that in the first test tube the level of the water meniscus has risen, while in the second it has fallen, although the distance through which the meniscus will move will not be as great in the second tube as in the first. In the first tube ether is diffusing through the water membrane more rapidly than chloroform, hence the volume of liquid below the membrane is constantly increasing, and the layer of water rises.

In the second test tube the chloroform diffuses through the water membrane more rapidly than the xylene hence the position of the layer of water is lowered. This behaviour is in accordance with the known solubilities of these three compounds in water; ether is the most soluble, chloroform next, and xylene the least. Evidently in this experiment permeability of the water membrane to these three compounds is directly correlated with their solubility in water.

The most familiar form of the solubility theory as applied to cytoplasmic membranes is the lipoid solubility theory first proposed by Overton.In brief this theory holds that substances soluble in lipoids pass readily through the cytoplasmic membranes while others do not. The term lipoid is used in a loose way to refer to fats and fat-like compounds. In recent usage the term lipid has been rather generally substituted for lipoid. Several lines of evidence appear to favour this theory.

Among these are:

- The non-miscibility of protoplasm with water, which probably indicates the presence of a film of fatty material on the outer surface of the protoplasm,
- The fact that most lipoids lower surface tension, and hence theoretically would accumulate at interfaces,
- The high resistance of protoplasm to the passage of an electrical current, which indicates that it contains layers through which ions do not diffuse readily. A fourth line of evidence for this theory comes from Overtons own experiments. He foundan almost perfect correlation between the lipoid solubility of hundreds oforganic compounds and the permeability of the cytoplasmic membranes to them. On the other hand, sugars, salts, and other lipoidinsoluble substa-nces showed practically no penetration into the cells at all. We now know that the terms lipoid-soluble and lipoid-insoluble correspond very closely to the modern terms of non-polar and polar, respectively.

While Overtons experiments seemed to show that the cytoplasmic membranes are relatively impermeable to polar compounds, later work has not all supported this generalization. It is true that polar compounds often penetrate with much greater difficulty than non-polar compounds and that the permeability of a given membrane to them may fluctuate, but the membranes are at times at least, permeable to them.

Moreover on a priori grounds it is evident that the membranes must be permeable to sugars, salts, etc. at least part of the time, since in the majority of cells most of these compounds present must have entered the cell at some time or another during its life history.Some investigators hold that the lipoid solubility theory is essentially a correct representation of the mechanism of cytoplasmic permeability.

Polar compounds would not be completely insoluble even in a lipoidal layer, and hence could penetrate through the membrane at relatively slow rates. Osterhout on siders that electrolytes must diffuse through this layer principally in the form of molecules as their dissociation when dissolved in a fat like liquid would be very slight. Other authorities are not convinced that this theory presents a complete picture of the basis for differential permeability in the cytoplasm.

Observed rates of penetration of some polar compounds seem too great to be accounted for satisfactorily by the lipoid-solubility hypothesis. Modifications of this theory have therefore been suggested. For instance Nathansohn has proposed a theory which postulates a membrane composed of a mosaic of proteins and lipoids. He assumed that polar (lipoid-insoluble) compounds could diffuse through the more or less hydrated protein portion of the membrane. Permeability to the latter class of compounds might vary with changes in the hydration of the proteins. Although this view is ahriost purely speculative some properties of the membrane, such as its elasticity, suggest the presence of proteins as one of its constituents.

ADSORPTION AND CHEMICAL REACTION THEORIES

It has been suggested that the permeability to non-polar compounds is correlated more closely withthe adsorptive capacity of the particles of the membrane for them than to their solubility in membrane constituents. In brief this adsorption theory would hold that the more strongly molecules are adsorbed, the more readily they will pass through the membrane.

The possible mechanics of the transport of adsorbed molecules across a membrane is described later. Chemical reactions may be involved in the transport of certain types of solutes across the cytoplasmic membranes. A penetrating molecule or ion according to such a chemical reaction theory, is supposed to combine with some substance at the outer boundary of the membrane, thus forming an intermediate compound, in the form of which it is presumed to move across the membrane.

Once the other boundary of the membrane has been reached a second chemical reaction is supposed to occur resulting in release of the ion or molecule on the interior of the membrane. Whether the penetration of a molecule depends upon solution, adsorption or chemical reaction the probability is strong that the cytoplasm itself actively participates in some manner in the movement of substances through it. For example, after molecules of a penetrating substance dissolve in, are adsorbed by, or combine chemically with micelles of the membrane, their penetration may be facilitated by kinetic activity of the membrane particles. A micelle might act in the capacity of a ferry boat, transporting a cargo of one or more molecules from one side of the membrane to the other Once across the membrane many of the passenger molecules would escape because of their relatively low concentration in the medium in contact with that side of the membrane.

A seemingly more probable hypothesis is that the cargoes of adsorbed, dissolved, or chemically bound molecules are not carried all the way across the membrane by one micelle, but pass from particle to particle as impacts occur between them. Each micelle would, according to this hypothesis carry a given molecule only for a very short portion of its journey across themembrane.

THE MECHANISM OF CHANGES IN MEMBRANE PERMEABILITY

The permeability of the cytoplasmic membranes fluctuates continuously as a normal feature of cellular activity. Some changes in the permeability of these membranes are induced by variations in factors of external origin as previously discussed, but others are due to unrecognized factors in the internal environment of the membranes. Furthermore, the living membrane of a given cell may at times be more permeable to some substances in certain regions thanin others, whereas a short time later the distribution of the relatively permeable are as may be very different. Various suggestions regarding the mechanisms by which changes in the permeability of the cytoplasmic membranes can take place have been proposed. .

Among these are:

- Changes in the viscosity of the membrane,
- Reversal of phase (on the assumption that the memb-rane has an emulsion-like structure), and
- Changes in the proportion of water in different phases of the colloidal system which is supposed to constitute the membrane. Since all of these hypotheses are highly speculative any detailed consideration of them in an introductory discussion would be unjustified.

6

Technology

Fruit quality goes back to tree stock, growing practices and weather conditions. Closer to the shipper and processor, however, are the degrees of maturity and ripeness when picked and the method of picking or harvesting. There is a distinction between maturity and ripeness of a fruit. Maturity is the condition when the fruit is ready to eat or if picked will become ready to eat after further ripening. Ripeness is that optimum condition when colour, flavour and texture have developed to their peak. Some fruit is picked when it are mature but not yet ripe. This is especially true of very soft fruit like cherries and peaches, which when fully ripe are so soft as to be damaged by the act of picking itself. Further, since many types of fruit continue to ripen off the tree, unless they were to be processed quickly, some would become overripe before they could be utilised if picked at peak ripeness. From a technological point of view, fruit characterisation by species and varieties is performed on the basis of physical as well chemical properties: shape, size, texture, flavour, colour/pigmentation, dry matter content (soluble solids content), pectic substances, acidity, vitamins, etc.

These properties are directly correlated with fruit utilisation. The proper time to pick fruit depends upon several factors; these include variety, location, weather, ease of removal from the tree (which change with time), and purpose to which the fruit will be put. For example, oranges change with respect to both sugar and acid as they ripen on the tree; sugar increases and acid decreases. The ratio of sugar to acid determines the taste and acceptability of the fruit and the juice. For this reasons, in some countries there are laws that prohibit picking until a certain sugar-acid ratio has been reached. In the case of much fruit to be canned, on the other hand, fruit is picked before it is fully ripe for eating since canning will further soften the fruit.

QUALITY MEASUREMENTS

Many quality measurements can be made before a fruit crop is picked in order to determine if proper maturity or degree of ripeness has developed. Colour may be measured with instruments or by comparing the colour of fruit on the tree with standard picture charts. Texture may be measured by compression by hand or by simple type of plungers. As fruit mature on the tree its concentration of juice solids, which are mostly sugars, changes. The concentration of soluble solids in the juice can be estimated with a refractometre or a hydrometre. The refractometre measures the ability of a solution to bend or refract a light beam which is proportional to the solution's concentration. A hydrometre is a weighted spindle with a graduated neck which floats in the juice at a height related to the juice density.

The acid content of fruit changes with maturity and affects flavour. Acid concentration can be measured by a simple chemical titration on the fruit juice. But for many fruits the tartness and flavour are really affected by the ratio of sugar to acid. Percentage of soluble solids, which are largely sugars, is generally expressed in degrees Brix, which relates specific gravity of a solution to an equivalent concentration of pure sucrose. In describing the taste of tartness of several fruits and fruit juices, the term "sugar to acid ratio" or "Brix to acid ratio" are commonly used. The higher the Brix the greater the sugar concentration in the juice; the higher the "Brix to acid ratio" the sweeter and lees tart is the juice.

HARVESTING AND PREPROCESSING

Harvesting

The above and other measurements, plus experience, indicate when fruit is ready for harvesting and subsequent processing. A large amount of the harvesting of most fruit crops is still done by hand; this labour may represent about half of the cost of growing the fruit. Therefore, mechanical harvesting is currently one of the most active fields of research for the agricultural engineer, but also requires geneticists to breed fruit of nearly equal size, that matures uniformly and that is resistant to mechanical damage.

A correct manual harvesting includes some simple but essential rules:

- The fruit should be picked by hand and placed carefully in the harvesting basket; all future handling has to be performed carefully in order to avoid any mechanical damage;
- The harvesting basket and the hands of the harvester should be clean;
- The fruit should be picked when it is ready to be able to be processed into a quality product depending on the treatment which it will undergo.

It is worth emphasising the fact that the proximity of the processing centre to the source of supply for fresh raw materials presents major advantages; some are as follows:

- Possibility to pick at the best suitable moment;
- Reduction of losses by handling/transportation;
- Minimises raw material transport costs;

- Possibility to use simpler/cheaper receptacles for raw material transport.

Once it has left the tree, the organoleptic properties, nutritional value, safety and aesthetic appeal of the fruit deteriorates in varying degrees. The major causes of deterioration include the following:

- Growth and activity of micro-organisms;
- Activities of the natural food enzymes;
- Insects, parasites and rodents;
- Temperature, both heat and cold;
- Moisture and dryness;
- Air and in particular oxygen;
- Light and
- Time.

Reception—Quality and Quantity

Fruit reception at the processing centre is performed mainly for following purposes:

- Checking of sanitary and freshness status;
- Control of varieties and fruit wholeness;
- Evaluation maturity degree;
- Collection of data about quantities received in connection to the source of supply: outside growers/farmers, own farm.

Variety control is needed in order to identify that the fruit belongs to an accepted variety as not all are suitable for different technological processes. Fruit maturity degree is significant as industrial maturity is required for some processing/preservation methods while for others there is the need for an edible maturity when the fruit has full taste and flavour. Special attention is given to size, appearance and uniformity of fruit to be processed, mainly in the form of fruit preserved with sugar using whole/half fruits ("with fruit pieces"). Some laboratory control is also needed, even if it not easy to precisely establish the technological qualities of fruit because of the absence of enough reliable rapid analytical methods able to show eventual deterioration.

The only reliable method for evaluating the quality is the combination of data obtained through organoleptic/taste controls and by simple analytical checks which are possible to perform in a small laboratory: percentage of soluble solids by refractometre, consistency/texture measured with simple penetrometres, etc.

Temporary Storage before Processing

This step has to be as short as possible in order to avoid flavour losses, texture modification, weight losses and other deterioration that can take place over this period.

Some basic rules for this step are as follows:

- Keep products in the shade, without any possible direct contact with sunlight;
- Avoid dust as much as possible;
- Avoid excessive heat;
- Avoid any possible contamination;
- Store in a place protected from possible attack by rodents, insects, etc.

Cold storage is always highly preferred to ambient temperature. For this reason a very good manufacturing practice is to use a cool room for each processing centre; this is very useful for small and medium processing units as well.

Washing

Harvested fruit is washed to remove soil, micro-organisms and pesticide residues. Fruit washing is a mandatory processing step; it would be wise to eliminate spoiled fruit before washing in order to avoid the pollution of washing tools and/or equipment and the contamination of fruit during washing. Washing efficiency can me gauged by the total number of micro-organisms present on fruit surface before and after washing - best result are when there is a six fold reduction. The water from the final wash should be free from moulds and yeast; a small quantity of bacteria is acceptable. Fruit washing can be carried out by immersion, by spray/showers or by combination of these two processes which is generally the best solution: pre-washing and washing.

Some usual practices in fruit washing are:

- Addition of detergents or 1.5 per cent HCl solution in washing water to remove traces of insect-fungicides;
- Use of warm water (about 50°C) in the pre-washing phase;
- Higher water pressure in spray/shower washers.

Washing must be done before the fruit is cut in order to avoid losing high nutritive value soluble substances (vitamins, minerals, sugars, etc.).

Sorting

Fruit sorting covers two main separate processing operations:

- Removal of damaged fruit and any foreign bodies (which might have been left behind after washing);
- Qualitative sorting based on organoleptic criteria and maturity stage.

Mechanical sorting for size is usually not done at the preliminary stage. The most important initial sorting is for variety and maturity. However, for some fruit and in special processing technologies it is advisable to proceed to a manual dimensional sorting (grading).

Trimming and Peeling (Skin Removal)

This processing step aims at removing the parts of the fruit which are either not edible or difficult to digest especially the skin. Up to now the industrial peeling of fruit and vegetables was performed by three procedures:

- Mechanically;
- By using water steam;
- Chemically; this method consists in treating fruit and vegetables by dipping them in a caustic soda solution at a temperature of 90 to 100° C; the concentration of this solution as well as the dipping or immersion time varying according to each specific case.

Cutting

This step is performed according to the specific requirements of the fruit processing technology.

Heat Blanching

Fruit is not usually heat blanched because of the damage from the heat and the associated sogginess and juice loss after thawing. Instead, chemicals are commonly used without heat to inactivate the oxidative enzymes or to act as antioxidants and they are combined with other treatments.

Ascorbic/Citric Acid Dip

Ascorbic acid or vitamin C minimises fruit oxidation primarily by acting as an antioxidant and itself becoming oxidised in preference to catechol-tannin compounds. Ascorbic acid is frequently used by being dissolved in water, sugar syrup or in citric acid solutions. It has been found that increased acidity also helps retard oxidative colour changes and so ascorbic acid plus citric acid may be used together. Citric acid further reacts with (chelates) metal ions thus removing these catalysts of oxidation from the system.

Sulphur Dioxide Treatment

Sulphur dioxide may function in several ways:

- Sulphur dioxide is an enzyme poison against common oxidising enzymes;
- It also has antioxidant properties; *i.e.*, it is an oxygen acceptor (as is ascorbic acid);
- Further SO_2 minimises non enzymatic Maillard type browning by reacting with aldehyde groups of sugars so that they are no longer free to combine with amino acids;
- Sulphur dioxide also interferes with microbial growth.

In many fruit processing pre-treatments two factors must be considered:

- Sulphur dioxide must be given time to penetrate the fruit tissues;
- SO_2 must not be used in excess because it has a characteristic unpleasant taste and odour, and international food laws limit the SO_2

content of fruit products, especially of those which are consumer oriented (*e.g.* except semi-processed products oriented to further industrial utilisation).

Commonly a 0.25 per cent solution (except for semi-processed fruit products which are industry oriented and use a 6 per cent solution) of SO_2 or its SO_2 equivalent in the form of solutions of sodium sulphite, sodium bisulphite or sodium/potassium metabisulphite are used. Fruit slices are dipped in the solution for about two to three minutes and then removed so as not to absorb too much SO_2.

Then the slices are allowed to stand for about one to two hours so that the SO_2 may penetrate throughout the tissues before processing. Sulphur dioxide is also used in fruit juice production to minimise oxidative changes where relatively low heat treatment is employed so as not to damage delicate juice flavour.

Dry sulphuring is the technological step where fruit is exposed to fumes of SO_2 from burning sulphur or from compressed gas cylinders; this treatment could be used in the preparation of fruits (and some vegetables) prior to drying/dehydration.

Sugar Syrup

Sugar syrup addition is one of the oldest methods of minimising oxidation. It was used long before the causative reactions were understood and remains today a common practice for this purpose. Sugar syrup minimises oxidation by coating the fruit and thereby preventing contact with atmospheric oxygen. Sugar syrup also offers some protection against loss of volatile fruit esters and it contributes sweet taste to otherwise tart fruits. It is common today to dissolve ascorbic acid and citric acid in the sugar syrup for added effect or to include sugar syrup after an SO_2 treatment.

Fresh Fruit Storage

Some fruit species and specially apples and pears can be stored in fresh state during cold season in some countries' climatic conditions. Fruit for fresh storage have to be autumn or winter varieties and be harvested before they are fully mature. This fruit also has to be sound and without any bruising; control and sorting by quality are mandatory operations.

Sorting has to be carried out according to size and weight and also by appearance; fruit which is not up to standard for storage will be used for semi-processed product manufacturing which will be submitted further to industrial processing. Harvested fruit has to be transported as soon as possible to storage areas. Leaving fruit in bulk in order to generate transpiration is a bad practice as this reduces storage time and accelerates maturation processes during storage. In order to store large quantities of fruit, silos have to be built.

Processing of Fruit Bars

The fruit bar processing method developed for FAO only involves a single major operation, which it drying the fruit pulp after mixing it with suitable ingredients. It can be used to produce mango, banana, guava or mixed fruit bars.

A dual-powered dryer, working by solar energy during the day and by electric or steam power at night and on rainy days, with cross-flow movement of air and controlled temperature (from 55° C at the beginning of processing to a high of 70° C), is well suited for dehydration of the pulp to the desired moisture level of 15 to 20 per cent.

Mango fruit bar — Fully ripe mangoes are selected and washed in water at room temperature. The peeled fruit is cut into slices and passed through a helicoidal pulper to extract the pulp. The required amount of sugar to adjust the Brix (the unit measure for total solids in fruits) of the mixed pulp to 25 degrees Brix is then added.

Two grams of citric acid per kilogram of pulp (or 20 ml of lime or lemon juice) are added to inhibit possible growth of micro-organisms during drying. The mixture is then heated for two minutes at 80° C and partially cooled; the heat treatment serves to inactivate the enzymes and destroy the micro-organisms. Potassium or sodium metabisulphite is added (two grams per kg of prepared mixture), so that the concentration of SO2 is 1000 ppm. The mixture is then transferred to stainless steel trays which have been previously smeared with glycerine (40 ml/m^2). Each tray must be loaded with 12.5 kg of mixture per square metre. Drying could be carried out by a dual-powered dryer for a total of 26 hours:

- 10 hours by solar energy at about 55° C and
- 16 hours by electric or steam power at 70° C.

At the end of the drying operation, when moisture content is between 15 and 20 per cent, the pieces of suitable shape and size are wrapped in cellophane paper, packed in cartons and stored at ambient air temperature.

Pieces of unsuitable shape and size are further cut into small pieces and used to prepare, along with peanuts and cashews, a variety of "cocktail mixtures". Banana fruit bar. - Banana varieties which give smooth pulp without serum separation must be used for this purpose.

Ripe, suitable fruit is selected. The hand-peeled fruits are soaked in 0.3 per cent citric acid solution for about 10 minutes (lime or lemon juice can replace citric acid). The drained fruit are pulped to obtain smooth pulp. The rest of the procedure is the same as in the case of the mango bar.

Guava Fruit Bar

A mixture of pink and yellow varieties is best suited for preparing the bar. The washed fruit is hand peeled and stem and blossom ends trimmed. The peeled fruit is cut into quarters which are passed through a helicoidal extractor to separate seeds and fibrous pieces (the holes in the stainless steel screen should be between 0.8 and 1.10 mm). To get the maximum yield of pulp, the material is passed through the extractor twice. After adjusting the refractometric solids to 25 degrees Brix, the fruit bar can be prepared by following the same procedure as for mango pulp.

Mixed Fruit Bar

Mango and banana pulp, as well as papaya and banana pulp, can be mixed in a calculated ratio for preparing mixed fruit bars. The rest of the procedure is the same as in the case of pure mango pulp.

Packing and Storage

The dried pulp is removed from the dryer and cut into square pieces of 5 x 5 cm at a thickness of about 0.3 cm. These pieces, arranged in three layers make up blocks of about 0.9 cm thickness weighing between 25 and 28 grams. An unit pack consist of two such blocks and weights between 50 and 56 grams. Each block is separately wrapped in cellophane and the unit pack is filled in a printed cellophane bag of size 15 x 8 cm. Two hundred unit packs are packed in a master carton of size 34 x 22 x 14 cm, with a net weight of about 10 kg. Shelf-life is about one year at room temperature.

Fruit Leathers

Fruit leathers are manufactured by drying/dehydration of fruit purées into leathery sheets. The leathers are eaten as confections or cooked as a sauce. They are made from a wide variety of fruits, the more common being apple, apricot, banana, cherry, blackcurrant, grape, peach, pear, pineapple, plum, raspberry, strawberry, kiwi fruit, mango and papaya.

A description of procedures for mango, banana, guava and mixed fruit bars is given in this document. Another product with good potential is ciku leather; ciku fruit is grown in Malaysia. A standard process is carried out using ripe fruits which are washed, peeled, diced and the seeds removed.

The fruits are blanched for 1 minute at 80° C and blended into puree in a food processor. Ciku leather is prepared by mixing ciku puree with 10 per cent sugar, 10 per cent pre-gelatinous rice flour, 150 ppm sorbic acid an 500 ppm sodium metabisulphite ($Na2H2SO4$). The mixture is cooked on a water bath at 60° C and then made into sheets 1.8 mm thick on trays spread with glycerol to reduce stickiness.

This is then further dried in a forced-air dehydrator at 45° C for 3.5 hr or until the surface no longer feels sticky when touched with the fingers. The dried and cooled leathers are cut into 12 x 12 cm squares and wrapped in polypropylene (PP) of 0.1 mm thickness.

Osmotic Dehydration in Fruit and Vegetable Processing

Osmotic dehydration is a useful technique for the concentration of fruit and vegetables, realised by placing the solid food, whole or in pieces, in sugars or salts aqueous solutions of high osmotic pressure. It gives rise to at least two major simultaneous counter-current flows: a significant water flow out of the food into the solution and a transfer of solute from the solution into the food.

Process Variables

Main process variables are:

- Pre-treatments;
- Temperature;
- Nature and concentration of the dehydration solutions;
- Agitation;
- Additives.

In the light of the published literature, some general rules can be noted:

- Water loss and solid gain are mainly controlled by the raw material characteristics and are certainly influenced by the possible pre-treatments;
- It is usually not worthwhile to use osmotic dehydration for more than a 50 per cent weight reduction because of the decrease in the osmosis rate over time. Water loss mainly occurs during the first 2 hr and the maximum solid gain within 30 min.;
- The rate of mass exchanges increases with temperature but above 45 ° C enzymatic browning and flavour deterioration begin to take place. High temperatures, *i.e.* over 60° C, modify the tissue characteristics so favouring impregnation phenomena and thus the solid gain;
- The best processing temperature depends on the food; mass exchanges are favoured by using high concentration solutions;
- Phenomena which modify the tissue permeability, such as over-ripeness, pre-treatments with chemicals (SO_2), blanching or freezing, favour the solid gain compared to water loss because impregnation phenomena are enhanced;
- The kind of sugars utilised as osmotic substances strongly affects the kinetics of water removal, the solid gain and the equilibrium water content. Low molar mass saccharides (glucose, fructose, sorbitol, etc.) favour the sugar uptake;
- Addition of NaCl to osmotic solutions increases the driving force for drying. Synergistic effects between sugar and salt have also been observed.

Applications

The effects of osmotic dehydration as a pre-treatment are mainly related to the improvement of some nutritional, organoleptic and functional properties of the product. As osmotic dehydration is effective at ambient temperature, heat damage to colour and flavour is minimised and the high concentration of the sugar surrounding fruit and vegetable pieces prevents discolouration. Furthermore, through the selective enrichment in soluble solids high quality fruit and vegetables are obtained with functional properties "compatible" with different food systems. These effects are obtained with a reduced energy input over traditional drying process. The main energy-consuming step is the reconstitution of the diluted osmotic solution that could be obtained by concentration or by addition of sugar.

Drying

Air drying following osmotic dipping is commonly used in tropical countries for the production of so-called "semi-candied" dried fruits. The sugar uptake, owing to the protective action of the saccharides, limits or avoids the use of SO2 and increases the stability of pigments during processing and subsequent storage period.

The organoleptic qualities of the end product could also be improved because some of the acids are removed from the fruit during the osmotic bath, so a blander and sweeter product than ordinary dried fruits is obtained. Owing to weight and volume reduction, loading of the dryer can be increased 2-3 times. The combination of osmosis with solar drying has been put forward, mainly for tropical fruit. A 24 hour cycle has been suggested combining osmodehydration, performed during the night, with solar drying during the day. Two-three-fold increase in the throughput of typical solar dryers is feasible, while enhancing the nutritional and organoleptic quality of the fruits. A two-step drying process, OSMOVAC, for producing low moisture fruit products was described. The osmotic step is performed with sucrose syrup 65-75 Brix until the weight reduction reaches 30-50 per cent. By osmotic dehydration followed by vacuum drying puffy products with a crisp, honeycomb-like texture can be obtained at a cost comparatively lower than freeze-drying. Commercial feasibility of the process on bananas has been studied, based on the results of a semi-pilot scale operation; the process scheme. Osmotically dried bananas retained more puffiness and a crisper texture than simple vacuum dried ones, and the flavour lasted longer at ambient temperature. The combination of osmotic dehydration with freeze-drying has been proposed only at laboratory scale.

Appertisation

A combination of osmotic dehydration with appertisation has been proposed to improve canned fruit preserves. The feasibility of a process, called osmo-appertisation, to obtain high quality fruit in syrup, has been assessed on a pilot scale. The key point of this technique is the pre-concentration of the fruit to about 20-40 Brix, that causes, together with the enhancement of the natural flavour, an increase of the resistance of the fruit to the following heat treatment, especially for colour and texture stability. The products obtained are stable up to 12 months at ambient temperature and show a higher organoleptic quality than canned preserved alternatives. Furthermore, because of their higher specific weight and diminished volume, the filling capacity of jars or pouches is increased.

Freezing

The frozen fruit and vegetable industry uses much energy in order to freeze the large quantity of water present in fresh products. A reduction in moisture content of the material reduces refrigeration load during freezing. Other advantages of partially concentrating fruits and vegetables prior to freezing include savings in packaging and distribution costs and achieving higher product quality because of the marked reduction of structural collapse and dripping during thawing. The products obtained are termed "dehydro-frozen" and the concentration step is generally carried out through conventional air drying, the additional cost of which has to be taken into account. Osmotic dehydration could be used instead of air drying to obtain an energy saving or a quality improvement especially for fruit and vegetable sensitive to air drying.

Extraction of Juices

An osmotic pre-step before juice extraction was reported to give highly aromatic fruit or vegetable juice concentrates.

Further Developments

So far only applications on a pilot plant scale are reported in the literature. For further developments on a larger scale, theoretical and practical problems should be solved. The industrial application of the process faces engineering problems related to the movement of great volumes of concentrated sugar solutions and to equipment for continuous operations. The use of highly concentrated sugar solutions creates two major problems. The syrup's viscosity is so great that agitation is necessary in order to decrease the resistance to the mass transfer on the solution side. The difference in density between the solution (about 1.3 kg/litre) and fruit and vegetables (about 0.8 kg/litre), makes the product float. Another important aspect, so far not investigated, is the microbiological safety of the process, which should be studied thoroughly before further industrial development.

Apricot Processing

- *Fresh apricot puree:* After washing, cutting and removal of stones, apricot halves are dipped in 2 per cent solution of sodium or potassium metabisulphite for 10 minutes. After draining, the resulting material is passed through a 0.045-in. screen pulper - finisher to produce a fresh apricot puree.

The fresh apricot puree obtained in this way could be further processed in different semiprocessed (*i.e.* chemically or otherwise preserved products) or finished fruit products (fruit leathers, fruit bars, jams, etc.).

- *Concentrated apricot pulp:* Fresh apricot halves could also be steam blanched for 5 min., passed through a 0.045-in pulper - finisher and transformed in a purée with about 14 Brix depending on the fruit quality. This purée may be concentrated in steam jacketed kettles up to 20 Brix or in other adequate equipment (*e.g.* a stirred vacuum evaporator) up to 28°Brix.

  As for fresh apricot purée, the concentrate may be further processed in various semiprocessed or finished fruit products.
- *Dried apricot leather*
  - From fresh fruit purée by drum drying. The fresh apricot purée at about 14 Brix could be dried to 12 per cent moisture apricot sheets, using a double-drum dryer operating at 132 degrees C with a drum clearance of 0.008 in and speed of 45 sec per revolution.
  - From fruit concentrate by drum drying. The concentrate could also be dried to 12 per cent moisture fruit sheets by the same process as described above.
  - From fresh fruit purée or from apricot concentrate by sun/solar drying or by dehydration.
- *Trays:* For sun/solar drying or dehydration of fruit pulp, the trays must have a solid base in order to retain the liquid contents. They may be made of metal, timber or plastic. Stainless steel or plastic trays are most suitable because they are unaffected by acid fruit pulp; they are, however, expensive.

  A metal tray could be 75 x 50 cm in size and with side 5 cm high. The trays must keep level during drying; if the tray is not level the pulp will run to the lowest point, giving a layer of irregular depth which will dry unevenly. Any tray which is not made of stainless steel or plastic must be covered inside with a sheet of heavy gauge plastic film to protect the pulp from chemical or bacteriological contamination.

  Standard sun/solar trays as described can be used by covering them inside with a sheet of plastic film to create a solid base.
- *Preparation before drying/dehydration*: Fresh apricot purée can be directly used for next processing steps. Fruit concentrate needs to

be added to potassium metabisulphite to obtain a 0.3 per cent concentration of SO2 in the material.

- *Drying/dehydration*: The apricot/fruit purée or concentrate is poured into the trays to a depth of about 1.5 cm. When stainless steel or plastic trays are used they should be coated with a thin layer of glycerine to prevent sticking.

The pulp is then sun/solar dried or tunnel/cabinet dehydrated; moisture content in the dried product should not exceed 14 per cent and the SO_2 content should not be less than 1500 pp. The dried product is wrapped in cellophane to prevent sticking, then put inside polythene bags and stored at best in tight fitting tins and sealed to prevent moisture transfer. From fresh fruit purée or from apricot concentrate, with sugar addition, and then processed by sun/solar drying or by dehydration.

In some countries preference is for finished products with added sugar; and this is also interesting from a point of view of energy consumption (concentration is partially achieved by sugar dry matter) and of shelf life. The overall content in SO_2 could also be reduced as sugar is a preservation agent, the product will be close to a fruit "paste".

Reconstitution Test for Dried/Dehydrated Products

In reconstitution water is added to the product which is restored to a condition similar to that when it was fresh. This enables the food product to be cooked as if the person was using fresh fruit or vegetable. All vegetables are cooked but many of the dried fruits can be used for eating after they have been soaked in water. The following reconstitution test is used to find out the quality of the dried product.

Reconstitution test:

1. Weigh out a sample of 35 grams from the bulked and packed final product of the previous day's production.
2. Put the sample into a small container (beaker) and add 275 ml of cold water (and 3.5 g salt).
3. Cover the container (with a watch-glass) and bring the water to the boil.
4. Boil GENTLY for 30 minutes.
5. Turn out the sample onto a white dish.
6. At least two people should then examine the sample for palatability, toughness, flavour and presence or absence of bad flavours. The testers should record their results independently.
7. The liquid left in the container should be examined for traces of sand/soil and other foreign matter.

This test can be used also to examine dried products after they have been stored for some time. Evaluation of rehydration ratio may be performed according to the following calculations. Rehydration ratio. If weight of the dried sample is 10 g (Wd) and the weight of the sample after rehydration is 60 g (Wr), rehydration ratio is:

$$\frac{Wr}{Wd} = \frac{60}{10} = \frac{6}{1}, 6 \text{ to } 1$$

Rehydration coefficient. The weight of rehydrated sample is 60 g (Wr); the weight of dried sample is 10 g (Wd) and its moisture is 5 per cent (Wu); raw material before drying had 87 per cent water (A); rehydration coefficient is:

$$\frac{Wr}{\frac{Wd - Wu \times 100}{100 - A}} = \frac{60 \times (100-87)}{10 - [10 \times 0.05]} = \frac{780}{9.5} = 52.1.$$

A simpler test for eating quality can be carried out without weighing and measuring. The material is placed in a cooking pot with water (and a little salt). The pot is then covered and boiled as described above. Except for a few products which are eaten in the dry state, most dried fruit and all dried vegetables are prepared by soaking and cooking. Often this preparation is carried out incorrectly and dried products get a bad reputation. Good quality dried products, after cooking and if properly treated should be similar to cooked fresh produce. In order to get good results, the following methods are recommended:

Quick Method

Cold water, ten times the weight of the dry product, is added to the dried product. The container is covered, brought to the boil and simmered GENTLY until the product is tender. The cooking time may be 15 to 45 minutes after the boiling point has been reached.

Slow Method

This gives better results than the quick method. Cold water is added to the dry food and is left to soak for 1 to 2 hours before cooking. The product is then cooked in the same water as that in which it was soaked. The actual cooking time will probably be shorter than that for the quick method.

Other points to remember are:

- If too much water is added the cooked product will have little flavour. However, if too little water is added the product may dry and burn. This can be avoided by adding small quantities of water during cooking;
- Always cook with a lid on the container;
- Salt, if required, should be added when the cooking is almost complete;
- Partly used packages of dry products should be reclosed tightly or kept in containers with good fitting lids.

Handling, Sorting, Packing and Storage of Dried and Dehydrated Fruit and Vegetables

It is not easy to assess when drying has been completed. In absence of instrumentation, the characteristics of the various products after drying/ dehydration can only be assessed by experience. Although this cannot be conveyed adequately on paper, some general indications can be given.

Fruit Products

When a handful of fruit is squeezed tightly together in the hand and then released, the individual pieces should drop apart readily and no moisture be left behind on the hand. It should not be possible to separate the skin by rubbing unpeeled fruit and the fruit centre should no longer reveal any moist area. Banana should be leathery and not too tough to eat in their dry state.

Vegetable Products

Onions should be dried until they are crisp whereas tomatoes should be leathery. In general, the lower the moisture content, the better the keeping quality will be, but overdried products generally have an inferior quality. Also the loss in weight from excessive drying cannot be tolerated in a commercial operation designed to run profitably.

It is, however, essential to dry up to an optimum/safe moisture level, related to the type of the product and his designed shelf life, and to avoid running the risk of the products becoming spoiled due to excess water content. When drying is completed, the material should be sorted either on trays or on a table in order to remove pieces of poor quality and colour and any foreign matter.

Very fine material should be separated from the bulk of the material by using a sieve (12 or 16 mesh per inch). Bad quality products which show poor colour need to be removed from the bulk of finished product. After selection and grading, dried products should be packed immediately, preferably in polythene bags which must be folded and closed/tied tightly. However, plastic bags are easily damaged and therefore they must be packed into cartons or jute sacks before they are transported.

Deterioration of Dried Fruit During Storage

Dried fruit must be considered as a relatively perishable commodity in the same category as cereals, pulses and similar stored products. It is subject to deterioration resulting from mould growth, insect and mite infestation and physical and chemical changes.

Mould Growth

When the moisture content of dried fruit is allowed to exceed the maximum permissible level for safe storage then mould growth may occur. The moisture levels applicable to various types of fruit; and it can be seen that the safe moisture levels for dried fruit are much higher than those for other similar commodities. At the present time, suitable field moisture metres for use with dried fruit are not readily available, and moisture determinations can only be satisfactorily carried out where laboratory facilities are available. Various species of drought resisting fungi may develop on dried fruit when the moisture content is just above the safe level, and a number of osmophilic yeasts are quite commonly associated with spoilage in dried fruit.

Many of the yeasts bring about fermentation with the production of lactic acid or alcohol, and yeasts are frequently present in wart-like crystalline growths which occur in fruit which has become "sugared". In very moist fruit mucoraceous fungi may predominate and are visible as white fluffy growths on and within the fruit.

Mite Infestation

Severe mite infestations are often associated with the growth of osmophilic yeasts in fermenting dried fruit products. Many of these mites are unable to complete their development in the absence of yeast. They have been reported as occurring on dried fruit, and particularly figs and prunes in Mediterranean countries. Such infestations are difficult to eradicate and affect consumer acceptance of the contaminated products.

Insect Infestation

Insect infestations may begin in the field before harvest, may continue during bulk storage after drying, and unless measures are taken to prevent it, may occur in the finished packaged product during storage prior to distribution and consumption.

Regular treatments of the stack of dried fruit with a suitable insecticide will be necessary as a routine to combat light insect infestations. Pyrethrins synergised with piperonyl butoxide are commonly used as a surface spray or as an aerosol fog for this purpose. Heavy infestations will require that the fruit be fumigated.

TECHNOLOGY OF SEMI-PROCESSED FRUIT PRODUCTS

The semi-processed fruit products are manufactured in order to be delivered to industry processing centres (in the fruit producing country itself or in importing countries) where they will be further manufactured in consumer oriented finished products: jams, jellies, syrups, fruits in syrup, etc.

In the practice of semi-processed fruit products and for the purpose of this document the following categories are defined:

a. Fruit "pulps": semi-processed products, not refined, obtained by mechanical treatment (or, less often, by thermal treatment) of fruit followed by their preservation. Either whole fruit, halves or big pieces are used which enables easy identification of the species. "Pulps" can be classified in boiled or non boiled (raw).

b. Fruit "purées-marks": semi-processed products obtained by thermal and mechanical treatment or, very rare, raw and then refined, operations by which all non-edible parts (cores, peels, etc.) are removed. "Purées-marks" are classified in boiled (the more usual case) and non boiled (raw).

c. Semi-processed juices: products obtained by cold pressure or very rare by other treatments (diffusion, extraction, etc.) followed by the preservation.

Technical Processes for Preservation of Semi-processed Fruit Products

Preservation can be achieved by chemical means, by freezing or by pasteurization. The choice of preservation process for each individual case is a function of the semi-processed product type and the shelf life needed.

Chemical Preservation

In many countries, in practice, this is carried out with sulphur dioxide, sodium benzoate, formic acid and, on a small scale, with sorbic acid and sorbates.Preservation with sulphur dioxide is a widespread process because of its advantages: universal antiseptic action and very economic application. The drawbacks of SO_2 are: SO_2 turn firms the texture of some fruit species (pomaces), desulphiting is not always complete and recolouring of red fruits is not always complete after desulphitation. Practical preservation dosage levels with SO_2 for about 12 months is 0.18-0.20 per cent SO_2 (with respect to the product to be preserved).

This level could be reduced to 0.09 per cent SO_2 for 3 months and to 0.12 per cent SO_2 for 6 months preservation. The preservation with sulphur dioxide is in use mainly for "pulps" and for "purées-marks". Chemical preservation can be performed from a practical point of view by the utilisation of 6 per cent SO_2 water solutions or by direct introduction of sulphur dioxide gas in the product (for "purées-marks"). The preparation of 6 per cent SO_2 solutions is done by bubbling the gas from cylinders in cold water; from a 50 kg SO_2 compressed gas cylinder results 830 l of 6 per cent SO_2 solution. These SO_2 solutions have to be stored in cool places, in closed receptacles and with periodic concentration control/check by titration or by density measurements approximate results. Preservation with sodium benzoate has the following advantages: it does not firm up the texture and does not modify fruit colour. The disadvantages are: it is not a universal antiseptic, its action needs an acid medium and the removal is partial. Sodium benzoate is in use for "pulps" and for "purées-marks" but less for fruit juices.

Practical dosage level for 12 months preservation is 0.18-0.20 per cent sodium benzoate, depending on the product to be preserved. Sodium benzoate is used as a solution in warm water; the dissolution water level has to be at maximum 10 per cent reported to semi-processed product weight. Formic acid preservation is performed mainly for semi-processed fruit juices at a dosage level of 0.2 per cent pure formic acid (100 per cent). Formic acid is an antiseptic effective against yeasts, does not influence colour of products and is easily removed by boiling.

Formic acid could be diluted with water in order to insure a homogeneous distribution in the product to be preserved; water has to be at maximum 5 per cent of the product weight. Because of a potential effect of pectic substance degradation, formic acid is less in use for "pulps" and "purées-marks" preservation. Sorbic acid used as potassium sorbate (easily water soluble) can be used for preservation of fruit semi-processed products at a dosage level of 0. 1 per cent maximum. Advantages of sorbates are: they are completely harmless and without any influence on the organoleptic properties of semi-processed fruit products.

Preservation by Pasteurization

As fruit has a low pH, preservation of semiprocessed fruit products could also be performed by pasteurization (heat treatment step at maximum temperature of 100°C), the length of this step varying with the size of the receptacles. The advantages of this type of treatment are: hygienic process, which assure a long term preservation; the disadvantages are: need for air tight receptacles, and pectic substances could begin to deteriorate if the thermal treatment is too long. Thermal preservation of fruit semi-processed products could also be done by a "selfpasteurization": very hot semi-processed products are filled into receptacles (*e.g.* metal cans) which are sealed and then inverted in order to sterilise the air which goes through the hot fruit mass.

Preservation by freezing

This is done on an industrial scale in some countries and can be done with or without sugar addition. The advantages of this process are: absence of added substances; very good preservation of quality of fruit constituents (pectic substances, vitamins, etc.) and good preservation of organoleptic properties (flavour, taste, colour). Freezing is done at about -20 to -30° C and storage at -10 to -18° C. Freezing is applied mainly to semi-processed fruit products aimed at very high quality and high cost finished products. Technological flow-sheet for semi-processed fruit "pulps": chemical preservation. Sorting is needed in order to remove sub-standard fruit (with moulds, with diseases, etc.) and all foreign bodies. Washing is obligatory in order to remove all impurities which cannot be eliminated at the processing step in finished products. Coring and

Cutting, mainly for pomace fruits, has as main objective a better utilisation of preservation "space" in receptacles and is not mandatory; this will be defined by customer/supplier agreements/standards.

This operation is preferably performed by mechanical means. Preservation is carried out with the 6 per cent SO_2 solution which is added to the prepared fruits (placed in bulk in receptacles) in the quantity needed to obtain the preservation dosage level. For a better/homogeneous preservative distribution, the initial 6 per cent SO_2 solution could be diluted with water; however, the diluted solution (which will be filled in receptacles) has to be at a dosage level of less than 10 per cent of the semi-processed product weight.

For some soft fruit, especially strawberries, preservation is done with a mix of 6 per cent SO_2 solution and calcium bisulphite solution (containing also 6 per cent SO_2). Preparation of calcium bisulphite solution is done by the introduction of 30 kg of CaO in 1 m^3 SO_2 solution and mixing up to clarification. The resulting solution is mixed with the initial 6 per cent SO_2 solution, generally in a 1:1 ratio, but the ratio can be adapted to the fresh fruit texture. Firming of soft fruit texture by this treatment is based on the formation of calcium pectate with pectic substances from fruit tissues.

In the case of sodium benzoate, formic acid or potassium sorbate, the dosage levels to be used are as indicated above with the rule that it is not allowed to add more that 10 per cent liquid in receptacles on the prepared fruits. Preservation by pasteurization or "self-pasteurization" will need as additional steps: a) boiling with a minimum water addition (maximum 10 per cent); b) filling of receptacles; c) hermetic closing followed by d) pasteurization or "self-pasteurization".

Processing of Pineapple-papaya Jam

The fruit should be prepared as per previous instructions. For pineapples, the ends are removed and discarded; the cores and outer parts of the fruits are also removed. The fruit cylinders obtained are pulped through a special extractor (Fitzpatrick communiting machine) equipped with a 0.40-in screen sieve; the pulp thus obtained is used for making jam. The papaya are prepared by hand-peeling the fruit; the fruit is then halved and the seeds removed. It is then pulped in the communiting machine using a 0.40-in screen sieve. When ginger root is used as flavouring, it is peeled and macerated in a Kenwood blender to a very fine consistency. A typical formula for a pineapple-papaya jam (50:50 ratio) with ginger flavouring is given as follows:

Pineapple pulp	25.0
Papaya pulp	25.0 pounds
Cane sugar	50.0
Apple pectin (150 grade)	6.0 ounces
Citric acid	6.4
Fresh ground ginger	7.5

Processing is carried out in the following way: The weighed fruit pulp is placed in a stainless steel steam-jacketed kettle and heated to about 110°F under constant stirring.

When the product reaches this temperature, the heat is turned off. The pectin (mixed in about ten times its weight with some of the weighed sugar), is then mixed into the fruit pulp, stirring constantly in order to prevent the pectin from clotting.

When the pectin has dissolved, the remainder of the sugar is added and dissolved completely in the mixture. The heat is then turned on and the jam mixture is stirred constantly until it starts boiling vigorously. During the remainder of the cooking, the product is stirred occasionally. Near the finishing point (approximately 221° F), the citric acid and the ginger (if it is used) are also added.

Determination of the finishing point is done by removing samples at intervals, cooling, and reading the soluble solids by means of a refractometre equipped with a Brix scale. After the jam reaches the proper Total Soluble Solids content, the heat is turned off and the surface scum/foam is removed. The jam then is quickly put into receptacles which have been cleaned and sterilised with boiling in water for 30 minutes.

The filling operation is done rapidly in order to prevent the temperature of the jam from falling below 190° F. After filling, sterilised lids (boiled for 30 minutes in water) are placed on the receptacles and they are then sealed. After this operation the receptacles are inverted for about 3 minutes to insure that the lids are sterilised. The receptacles are then placed upright. At this stage it is not necessary to do any further processing, therefore the receptacles are cooled in running cold water until they reach a temperature slightly above room temperature. They are then dried in air and labelled.

Evaluation of Finished Products

During production at medium/large scale, it is recommended that quality controls be performed during manufacturing. After ten weeks of storage at room temperature it is recommended that an examination of finished products be performed. The receptacles are opened and contents carefully emptied on to enamel trays without disturbing the formation of the jam. The empty cans (if metal cans were used) are then inspected for signs of corrosion. Factors other than flavour include colour, appearance, syrup separation, firmness and spreading quality. For flavour, jam is tested on pieces of bread. Samples are taken for measurement of pH (with a glass electrode pH metre) and Total Soluble Solids (with a refractometre equipped with a Brix scale). This evaluation enables to have a quality check during product shelf life and to obtain data needed for necessary improvements of future productions. For pineapple-papaya jam, products made with 30 per cent pineapple and 70 per cent papaya with added ginger has the highest score for flavour. The use of plain tin cans causes corrosion problems which is not the case when acid resistant lacquer cans are used.

Gelified Sugar Fruit Preserves

Technology of Fruit Jellies

Jellies are gelified products obtained by boiling fruit juices with sugar, with or without the addition of pectin and food acids. Jellies are usually manufactured from juices obtained from a single fruit species only, obtained by boiling in order to extract as much soluble pectin as possible.

Jellies have to be clear, shiny, transparent and with a colour specific to the fruit from which they are obtained. Once the product is removed from the glass receptacles where it was packed, jellies must keep their shape and gelification and not flow, without being sticky or of a too hard consistency. Technological flow-sheet for jellies manufacturing covers two categories of operations: those to obtain gelifying juices and those related to the manufacturing of jelly itself.

a. Production of gelifying juices: Washing and Sorting are carried out in usual conditions; Cutting is applied eventually only to pomaces (apples, quinces) and are limited to cutting in halves or quarters;
   Boiling is performed with water addition with 50-100 per cent water, needed for pectin extraction. The boiling time is 30-60 min., it should not be longer so as to avoid pectin degradation; at the same time boiling must not be too violent.
   Juice separation is carried out by a simple drain through metallic sieve or cloths; in these cases the yield is lower and the residue can be used for marmalade production. In bigger productions it juice separation by hydraulic press is preferred, yield being in these cases greater.
   Juice clarification is strictly necessary in order to obtain clear jellies. This step can be achieved by sedimentation during 24 hours or by filtration.
b. Manufacturing of jellies: Basic Recipe Seting is done starting with equal parts in weight of sugar and juice (for example 1000 g juice and 1000 g sugar). As final jelly has to contain about 60 per cent added sugar, weight of finished product must be of about 1600 g, by evaporation of about 400 g water. Boiling is carried out as following: juice is boiled up to removal of about half of the water that has to be evaporated, then the calculated sugar quantity is added gradually; the remainder of the water is evaporated until a concentration in soluble substances (refractometric extract) of 65-67 per cent is reached, in which is incorporated also the sugar from juice.During boiling it is

necessary to remove foam/scum formed. Product acidity must be brought to about 1 per cent (malic acid) corresponding to pH > 3. Any acid addition is performed always at the end of boiling. For juices rich in pectin, gelification will occur without pectin addition. If at the trial boiling test the gelification has not occurred, because of pectin absence, in this case 1-2 per cent powder pectin will be added by operating as indicated: pectin is mixed with 10-20 fold sugar quantity and is introduced directly in the partially evaporated juice and then boiling is conducted rapidly up to final point. Evaluation of final point is done not only by refractometry but also by gelifying test.

A rapid test for evaluation of juice pectin content is possible by mixing a small sample of juice with an equal volume of 96 per cent alcohol; the apparition of a compact gelatinous precipitate indicates a sufficient pectin content for gelificatioBoiling of jellies is performed in small batches (25-75 kg) in order to avoid excessively long boiling time which brings about pectin degradation.

Cooling is optional and is carried out up to 85 deg. C, in double wall baths with water circulation.

Filling is performed at a temperature not below 85 deg. C in receptacles (glass jars, etc.), which must be maintained still about 24 hours to allow cooling and product gelification.

Receptacle closing is done after product gelification.

Usual jelly types are: quinces, strawberries, cherries, wild berries, alone or in mixes with apples.

Grading of Marmalades

Three categories can be defined:

1. Fine marmalade, manufactured from one fruit;
2. Superior marmalade, obtained from a mix of fruit in which 30 per cent are "noble" species (cherries, strawberries, apricots, etc.) and 70 per cent from other species;
3. Marmalade from fruit mixes; apples, pears, plums, quinces, ungrafted apricots and wax cherries may be used, with the optional addition of "superior" fruit which was rejected at sorting but which was sound.

The content in total soluble substances (refractometric extract) of marmalades must be 64 per cent minimum; the acidity must be between 0.5 per cent and 1.8 per cent expressed as malic acid.

Basic Recipe Setting For a normal composition - marmalade without pectin addition the following is a basic recipe:

100 kg semi-processed fruit product
(10 per cent refractometric extract) .... 10 kg soluble substances

55 kg sugar .... 55 kg soluble substances
155 kg 65 kg soluble substances
55 kg water to be evaporated
100 kg marmalade with 65 per cent refractometric extract

This marmalade satisfies many standards and at same time has a good shelf-life since it contains less than 35 per cent water. Semi-processed fruit products must have a minimum 8 per cent refractometric extract; in this case the recipe should use 125 kg of raw material, with 80 kg water to be evaporate.

The use of semi-processed fruit products with a low refractometric extract presents the following drawbacks:

- Higher water quantity to be evaporated;
- Longer boiling times with negative impact of pectin degradation;
- Loss of flavour and
- Lower equipment efficiency.

Pectin addition in marmalade manufacture produces the following advantages:

- Improvement of gelification,
- Economy in fruit;
- Shorter boiling time; this maintains taste and flavour and produces higher equipment efficiency.

Pectin addition makes it possible to obtain the "fine" type of marmalade from "noble" fruits which do not contain enough pectin (cherries, peaches, apricots, etc.).

In marmalades from fruit mixes, low pectin content can be compensated by addition of semi-processed fruit products which are rich in this component (for example apples).

When pectin is to be added, the above recipe should be modified as follows:

80 kg semi-processed fruit product
(10 per cent refractometric extract) .... 8 kg soluble substances
55 kg sugar .... 55 kg soluble substances
10 kg pectic extract (10 per cent R.E.) .... 1 kg soluble substances
145 kg 64 kg soluble substances
45 kg water to be evaporated
100 kg marmalade with 64 per cent refractometric extract

Pectin can be added as pectic extract with about 10 per cent refractometric extract (R.E.) in the recommended proportion or in the form of a powder considered with 100 per cent dry matter (*e.g.* 100 per cent soluble substances) in a quantity of about 1 per cent.

In some countries it is usual to add 5-12 per cent of corn syrup (calculated to finished product weight) to replace the corresponding quantity of sugar (100 parts corn syrup can replace 80 parts sugar). Corn syrup has to be liquefied by heating before use. Corn syrup addition reduces the too sweet taste of marmalade, avoids sugar crystallisation and gives a special shine to finished products. Marmalade manufacturing follows the technological line and covers the following steps: "MARK" Preparation can be achieved from fresh fruits or starting from chemical preserved semiprocessed fruit products: "marks" or "pulps". In the latter case, pulps will be processed in marks which then will be desulphitated.

Desulphitation is carried out by boiling at atmospheric pressure, under vacuum in specialised equipment or under pressure in special retorts built in acid-resistant material. In any case, the desulphitation must be carried out before sugar addition because sugar will bound to the sulphur dioxide. The desulphitation operation must be conducted so as to be, if possible, fully completed; the finished product must contain less than 0.005 per cent free SO_2. The technological flow for marmalade production is the following: fresh fruit after sorting on control belt (1) is washed in a washing machine (2) is brought to the continuous boiling equipment (3) then to the pulper (4); the semi-finished mark is passed on to the storage tank (6). Pulps are boiled and desulphited in continuous boiling equipment (3), then are brought to pulper (4) and to the storage tank (6). Boiling aims at evaporating the required water quantity, to facilitate the formation of pectin-sugar-acid gel and to partially invert sugar (about 40 per cent from total sugar). The boiling operation can be carried out in open kettles or in evaporators under vacuum. In the latter case, the warm "mark" from storage tank (6) is aspirated in a concentrator (7) in a vacuum and submitted to a partial boiling up to removal of half of water quantity which needs to be evaporated; the calculated sugar quantity is then added by aspiration, keeping the boiling on.

After this the pectin extract or powder pectin which has previously been dissolved in warm water, is added; when the final concentration is reached, as indicated by refractometric control, the required quantity of acid is added. Sugar is added in proportion of 55 per cent in finished product, pectic extract (10 per cent refractometric extract) at a level of about 10-15 per cent and the acid (citric, tartaric, lactic) in a quantity needed to obtain a finished product acidity of about 1 per cent. Boiling at atmospheric pressure affects not only the appearance but also the nutritional value of the products, mainly if these contain proteins, as some albuminoids coagulate even at 60°C. Food products for which flavour is an essential property as for example fruit juices, etc., are also affected by the action of heat. Heat treatment has an impact on vitamin losses, mainly of vitamin C, in the presence of oxygen as is the case at concentration in open vessels. Sugars are generally less damaged by heat at temperatures below 100° C; as the boiling point is increasing above 100° C, a risk of partial sugar caramelization exists. In the study of heat effects on products submitted to concentration operation, it is necessary to take into account not only the evaporation surface temperature but also the distribution of the temperature in the whole liquid mass.

The length of the heating period also has a major influence because in many cases it is preferable to concentrate the liquid at a relatively high temperature in a short time avoiding the drawbacks of lower temperatures acting during a long time. In order to maintain the food value and organoleptic properties, it is necessary that concentration take place at a low temperature which can be achieved by concentration under partial vacuum, taking into account that boiling point decreases when the residual pressure decreases, respectively with the increase of vacuum degree.

Advantage of concentration under partial vacuum are the following:

- Lowering of boiling point;
- The total time needed for concentration of food products under a residual pressure of about 200 mm Hg is about half as compared to the that of concentration by boiling at atmospheric pressure;
- By lowering the concentration temperature and time, organoleptic properties and of nutritional value are maintained better particularly as far as the vitamins are concerned;
- When products are concentrated in a vacuum, it is possible to recover volatile aromatic substances by using adequate installations.

Technical procedures of concentration by vaporisation can be classified in:

- Concentration at atmospheric pressure: continuous or discontinuous;
- Concentration under partial vacuum: discontinuous (in vacuum equipment with simple or multiple effect) or continuous (in vacuum installations with continuous action or in thin film vaporisation installations).

Even if open kettle equipment is less expensive than evaporators in a vacuum, it is necessary to take into account that boiling under vacuum has the following advantages:

- Low boiling temperature (60-70° C), depending the degree of the vacuum; this give the fruit better taste and flavour-keeping qualities;
- Easy feeding with raw and auxiliary materials;
- Shorter boiling time;
- Better working conditions (vapour elimination in condensed water and not in open air).

There are small size evaporators under vacuum which can be well suited to the needs of medium size operations in developing countries. Cooling of marmalade to about 50-60°C can even be done in a vacuum evaporator by closing the heating steam and maintaining vacuum degree or by discharge in storage tanks (8). Filling in receptacles (boxes, jars, glasses, etc.) is done preferably with filling machines (9) followed by labelling(10). Small packages can be closed warm or after complete cooling; big packages (boxes, etc.) must be closed only after cooling, *e.g.* 24 hours after processing. Storage of marmalade must be done in dry rooms (air relative humidity at about 75 per cent), well ventilated, medium cool places (temperature 10-20 degrees C), disinfected and away from direct sunlight and heat. These measures are necessary because marmalade is a hydroscopic product and, by water absorption, favourable conditions for mould development are created.

Technology of Special Fruit Jams

Special fruit jams are products similar to marmalades but in which fruit partially keep their shape (whole, halves, etc.). Special fruit jams are obtained by boiling fruit with sugar, with or without pectin addition, with acid addition followed by n concentration by evaporation. Special fruit jams present a pronounced gelification at their cooling and can be considered as fruits included in a pectin-sugar-acid gel. High quality special fruit jams are obtained only from fresh fruit or possibly frozen and from only one fruit species.

Special fruit jams are classified in:

- Non-pasteurized (min. 68 per cent refractometric extract) and
- Pasteurized (min. 65 per cent refractometric extract); minimum acidity, expressed in malic acid, is 0.5 per cent.

Basic Recipe Seting is done taking into account following elements: - maintenance, as much as possible, of fruit shape, because this is specific to these finished products;

- Added sugar, in relation to finished product, must be at 60-65 per cent; this high percentage is needed in order to obtain a rapid and strong gelification and to facilitate conservation of fruit shape;
- A satisfactory gelification cannot generally be obtained without pectin addition, at a level of 1-2 per cent, or respectively 10-20 per cent pectic extract (10 per cent refractometric extract);
- Partial (about 40 per cent) inversion of added sugar is necessary.

The basic recipe is: 80 kg fruit + 100 kg sugar + 1.6-2.0 kg pectin powder and 1 kg citric or tartric acid; this will yield about 165 kg of special fruit jam with 60 per cent added sugar; water quantity to be evaporated is about 18 kg. If the finished product has to contain 65 per cent added sugar, the boiling has to be continued up to evaporation of about 30 kg water, the resulting finished product quantity will be about 153 kg. Technological flow sheet for manufacturing of special fruit jams is as follows: Fruit Preparation: sorting, washing, peeling and coring (for apples, pears, quinces), or removal of quetches and stones/pits (for plums, peaches, apricots, cherries) or of short tails (for strawberries and wild berries). Pomace fruits are then cut in slices or quarters. Boiling with sugar is the most important operation in production of special fruit jams and has as objective to evaporate water until gel formation and partial inversion of sugar. Boiling has to be conducted in such a way as to avoid fruit disintegration, but fruits must be well penetrated with sugar.By boiling an equilibrium is reached, by osmosis, between sugar solution and cellular juice. The initial concentration gap between sugar syrup and cellular juice is very high and if the equilibration process is forced, juice comes out of cells and fruit loses its shape and may even disintegrate. The boiling process accelerates the equilibration, intensity of which increases with temperature and boiling time. Pectin addition shortens boiling and thus delays the equilibration; for this reason there are different methods for special fruit jams preparation:

- Maintaining fruit in sugar, at ambient temperature, over 8-24 hours; fruit to sugar ratio is that indicated in recipe. After this sugar impregnation, fruits and resulting sugar syrup are brought together to boil; towards the end of boiling, pectin, dissolved in warm water, and necessary acid quantity are added.
- Preparation of a very concentrated sugar syrup (at least 75 per cent), in which fruit is introduced in order to be boiled. The rest of the operations are similar to those described in the previous method.
  - Soft fruit (strawberries, wild berries) can be mixed with sugar directly in evaporating open kettles, without added water and then heated gradually up to boiling, which is continued as in previous method.

Boiling is preferably carried out in small open kettles (50-100 kg) in order to avoid too long a boiling and fruit disintegration. Gelification corresponds generally to the reaching of a concentration of 65 per cent soluble extract, respectively 68 per cent refractometric extract. Practical test for gelification is done as for jellies and marmalades.

Cooling is a technological step strictly necessary in order to avoid fruit rising to the surface in preservation receptacles. Cooling is done in a double bottomed water bath in which water circulates at about 80° C. Filling of receptacles (jars, boxes, glasses, etc.) is carried out and it is necessary to assure at this stage that the finished product is homogeneous (equal quantities of fruits and gel). Pasteurization is only applied to special fruit jams with 65 per cent refractometric extract packed in jars or boxes and is performed at about 100° C for about 20 min. Gelification is carried out during product cooling and intensifies during storage. Storage must follow the same conditions as for marmalades.

NON- EELIFIED SUGAR FRUIT PRESERVES

TECHNOLOGY OF FRUIT JAMS

Fruit jams are products obtained by boiling of fruits (whole, halves, etc.) with sugar syrup until the reaching a viscous consistency. Jams can be defined as fruits included in a concentrated syrup. Jams are only prepared from fresh fruits; the usual product range covers the following species: strawberries, cherries, apricots, wild berries, peaches, plums, raisins, quinces, rose petals, etc.; manufacturing of jams from fruit mixes is not an accepted practice.

Standards indicate that fruit jams have to conform with following conditions:

- Fruit or fruit pieces content 45-55 per cent;
- Soluble substances, refractometric degree min. 72 per cent;
- Acidity (expressed in malic acid) min. 0.7 per cent.

Basic Recipe Setting is presented below as an indication:

- 100 kg fruits
- 150 kg sugar
- 250 kg
- 35 kg water to be evaporated
- 215 kg jam at about 72 per cent refractometric extract

Fruit preparation is similar to special fruit jams with the difference that stone/seed removal is mandatory for all species. Boiling Fruits with Sugar can be achieved by the same three methods described above for special fruit jams. By boiling a sugar content equilibration is foreseen and the operation must be conducted in such a way that texture, flavour and colour of fruits be preserved. Foam/scum has to be removed during boiling. Generally boiling in concentrated sugar syrup (at least 75 per cent) after a previous diffusion during 2-4 hours is in practical use. Boiling must to be carried out in small portions (about 15 kg) in order to avoid fruit disintegration.

For some fruit species, boiling has to be conducted in many phases/steps with "stops" to enable sugar diffusion. At the end of boiling, vanillin at a ratio of 125 g for 100 kg jam may be added for some fruits (white cherries, raisins, etc.). At the same time, it is also possible to add citric or tartric acid in order to avoid the "sugaring" defect. Cooling of jams, necessary in order to avoid fruit rising to the surface, is carried out as for special fruit jams. Filling of receptacles and Storage for jams are performed in same way as for special fruit jams.

Technology of Special non Gelified Fruit Marmalades

Special non gelified fruit marmalades are products resulting from fruit without stones or seeds, sieved or squashed, concentrated by boiling, without sugar added and Non- gelified. Their consistency results from a low water content (about 35 per cent) and a high percentage of insoluble substances (5-10 per cent). Sugar the from fruits acts as a preserving agent. Plum special Non- gelified fruit marmalade is the product representative of this category. Other fruit is used very rarely, because they have a reduced sugar content as compared to plums; though there are some countries producing special Non-gelified fruit marmalade from pears or sweet apples.

For plums, the finished product in this category must contain minimum 55-60 per cent soluble substances (refractometric extract), rising up to 70 per cent for a high quality product. Basoc Reco[e Settomg for plums is done in relation with sugar content of fruits; when this is higher, the product quality is better and the yield is higher. Thus, if plums with 18 per cent refractometric extract are used, 300-350 kg of fruits are needed for 100 kg finished product. At the same time, concentration of sugar by boiling, also increases by three fold acidity and astringency of this special plum Non- gelified marmalade; for this reason it is recommended to use only sweet and completely mature plums. Washing is performed in usual conditions. Pre-Boiling of fruits can be carried out in water or vapour, preferably with continuous running and has as its objective the softening of tissues. De-Stoning is performed in a pulper. Boiling of the sieved mass is done in double bottom open kettles with a big evaporation surface or in vacuum evaporators.

Boiling in open kettles enables production of a more tasty slightly caramelized, product; boiling in vacuum evaporators has the technological advantages indicated in marmalade production. At the end of boiling and once of necessary concentration is reached, the product is poured directly into receptacles (drums, etc.) and left to cool in order to form a hard surface layer (crust).

Storage of well closed receptacles is carried out as for marmalade. Special Non- gelified fruit marmalade can also be prepared from chemically preserved semi-processed fruit products, but the quality is lower than that obtained from fresh fruits. Sometimes dried prunes in a mix with preserved semi-processed products can be used for plum special Non- gelified marmalade preparation. In some countries plum finished products in this category are sweetened by the addition of maximum 30 per cent sugar, calculated in relation to the finished product.

Technology of Fruit Pastes

These products are obtained in a similar way to marmalades and special Non- gelified marmalades, but have a lower water content (about 25 per cent). Reduction of water content can be achieved by continuing the boiling of the product or by natural or artificial drying. A typical example of fruit paste without sugar added is the apricot paste - "pistil", etc. which is a concentrated special Non- gelified fruit marmalade poured in thin layers and sun dried.

An example of fruit paste with sugar added is quince paste which is a marmalade concentrated by evaporation. Sugar content must be 65 per cent; soluble substances content, 7075 per cent refractometric extract and acidity at least 0.5 per cent expressed as malic acid. Packing is done usually in polyethylene sheets and then in boxes or tins; storage conditions are similar to those for marmalade.

Technology of Fruit Syrups

Fruit syrups are products obtained by dissolving sugar in juices obtained from direct pressing of fruits. Sugar dissolving can be done at room temperature or by heating. Syrups have to contain 68 per cent soluble substances (refractometric extract) and minimum 1 g/100 ml malic acid. Up to a maximum 10 per cent of sugar can be replaced by corn syrup. Syrups must be manufactured from the juice of only one fruit species. Juice preparation is carried out at room temperature as described above.

Sugar dissolving in fruit juice can be performed by one of the following methods:

- Boiling in open kettles, done by using the basic recipe:
  350-400 kg fruit juice; 650-660 kg sugar; max. 10 kg citric or tartric acid. The juice is brought to boiling and the sugar is dissolved; the total time has to be as short as possible in order to avoid flavour loss and a too high sugar inversion degree (optimum inversion degree is 40 per cent). Acid is added preferably towards the end of boiling.
  During all boiling processes it is necessary to remove foam/scum. In order to avoid caramelization, the syrup has to be cooled rapidly, and this can be carried out in baths with double bottoms through which are circulated water.
  One alternative to this method is to boil syrup in closed vessels to avoid flavour losses.
- Boiling in a vacuum. The basic recipe is the same as above. Sugar and fruit juice are mixed previously in a pre-heating kettle and then transported to vacuum equipment.
  Boiling is performed at 50°C and at the end the temperature is raised slowly up to 65-70°C. The syrup can be cooled directly in vacuum equipment by closing the steam inlet and by increasing the vacuum. In this boiling method it is possible to incorporate a flavour recuperation device.
- A continuous process for syrup preparation can be carried out by dissolving components with heat while passing them through a horizontal cylinder with a screw inside.
  In the methods where sugar is dissolved by heat, it is also possible to use chemically preserved juices.
  In this case it is necessary to first perform the desulphitation of juices preserved with SO_2. This can be performed by boiling juice with optional water addition (and before any sugar addition). High quality syrups are obtained however from fresh juices.
- Sugar can also be dissolved at room temperature by using continuous flow percolators. These are similar to those used for salt solution preparation in vegetable canning processes. The juice goes over a sugar layer and is concentrated progressively until saturation (about 65 per cent). The syrup is then passed through a filtration section in the bottom of the percolator.

Syrup filtration is needed in order to clarify crystals; the filtration of syrup is done in warm conditions through cloth. Filling of syrup in bottles is done in aseptic conditions as much as possible in order to avoid syrup infection with osmophile yeasts. Syrup preservation is assured by the high sugar content with respect to a low water activity [unclear]. Storage takes place in well ventilated storage rooms; avoiding sunlight at 10-15°C. The usual product range is: strawberries, cherries, wild berries, citrus fruits.

TECHNOLOGICAL STEPS FOR PROCESSING OF FRUIT JUICES WITHOUT PULP

Fruit juices must be prepared from sound, mature fruit only. Soft fruit varieties such as grapes, tomatoes and peaches should only be transported in clean boxes which are free from mould and bits of rotten fruit. Washing: fruit must be thoroughly washed. Generally, fruit will be submitted to a pre-washing before sorting and a washing step just after sorting. Sorting: removal of partially or completely decayed fruit is the most important operation in the preparation of fruit for production of first quality fruit juices; sorting is carried out on moving inspection belts or sorting tables. Crushing/Grinding/Disinteration Step is applied in different ways and depends on fruit types:

- Crushing for grapes and berries;
- Grinding for apples, pears;
- Disintegration for tomatoes, peaches, mangoes, apricots.

This processing step will need specific equipment which differs from one type of operation to another. Enzyme Treatment of crushed fruit mass is applied to some fruits by adding 2-8 per cent pectolitic enzymes at about 50° C for 30 minutes.

This optional step has the following advantages: extraction yield will be improved, the juice colour is better fixed and finished product taste is improved. However, for fruit which is naturally rich in pectic substances, this treatment makes the resulting "exhausted" material useless for industrial pectin production. Heating of crushed fruit mass before juice extraction is an optional step used for some fruit in order to facilitate pressing and colour fixing; at same time, protein coagulation takes place.

Pressing to Extract Juice

Diffusion is an alternative step for juice extraction and can be carried out discontinuously or in batteries at water temperature of about 80-85 ° C. Juice clarifying can be performed by centrifugation or by enzyme treatment. Centrifugation achieves a separation of particles in suspension in the juice and can be considered as a pre-clarifying step. This operation is carried out in centrifugal separators with a speed of 6000 to 6500 RPM. Enzyme clarifying is based on pectic substance hydrolysis; this will decrease the juices' viscosity and facilitate their filtration. The treatment is the addition of pectolitic enzyme preparations in a quantity of 0.5 to 2 g/l and will last 2 to 6 hours at room temperature, or less than 2 hours at 50° C, a temperature that must not be exceeded.

The control of this operation is done by checking the decrease in juice viscosity. Sometimes, the enzyme clarifying is completed with the step called "sticking" by the addition of 5-8 g/hl of food grade gelatine which generates a flocculation of particles in suspension by the action of tannins. Filtration of clarified juice can be carried out with kieselgur and bentonite as filtration additive in press-filters (equipment).

De-Tartarisationis applied only to raisin juice and is aimed to eliminate potassium bitartrate from solution. This step can be performed by the addition of 1 per cent calcium lactate or 0.4 per cent calcium carbonate. Pasteurization of juice can be done for temporary preservation (pre-pasteurization) and in this case this operation is carried out with continuous equipment (heat exchangers, etc.); warm juice is stored in drums or large size receptacles (20-30 kg). Pasteurization conditions are at 75°C in continuous stream.

Pasteurization of bottled juice is then carried out just before delivery to the market; this is performed in water baths at 75° C until the point where the juice reaches 68° C. In cases when the final pasteurization is done without pre-pasteurization and temporary storage, modern methods use a rapid pasteurization followed by aseptic filling in receptacles. Rapid pasteurization conditions are as follows: temperature about 80° C, over 10-60 sec., followed by cooling; all operations are carried out in continuous stream.

Preservation under CO_2 pressure may be done at a concentration of 1.5 per cent CO_2 under a pressure of 7 kg/cm^2. At the distribution step, proceed at CO_2 decompression and the juice is then submitted to a sterilising filtration and aseptic filling in receptacles.

Preservation by freezing is carried out at about -30° C, after a preliminary de-aeration; storage is at -15 to -20° C. Production of concentrated juices by evaporation is performed under vacuum (less than 100 mm Hg residual pressure) up to a concentration of 65-70 per cent total sugar which assures preservation without further pasteurization. Modern evaporation installations recover flavours from juices which are then reincorporated in concentrated juices.

Additional operations for juice manufacturing are the vacuum de-aeration and mixing with other fruit juices or with sugar. For the production of Non-clarified juices the centrifugation is the only specific step, enzyme clarifying and subsequent filtration being eliminated. The optimum sugar/acid ratio for the majority of fruit, mainly for pomaces, is 10/1 to 15/1. Fruit which is rich in carotenoids (apricots, peaches, etc.) is only processed as juices with pulp ("nectars").

Technological Flow-sheet for Fruit Juices with Pulp ("Nectars")

This process is divided at industrial scale in two categories of operations:

1. Processing for obtaining juices;
2. Juice conditioning for preservation.

Pomaces (apples, pears) are washed and sorted and then crushed in a colloid mill; fruit purée is then passed through a screw type heating equipment where direct steam is used as a source of heat. Warm fruit mass is treated in a pulper with a 2 mm screen and then through an extractor similar with the equipment used for tomato juice. Stone fruits (apricots, peaches, cherries, etc.) after washing and sorting are submitted to steam in a continuous heater, then the warm fruit mass is passed through a pulper and then an extractor (as mentioned above). Berries (strawberry, wild berries, etc.) are washed, sorted and then crushed, preheated and then introduced in extractor. In order to avoid browning and undesirable taste modifications it is usual to add about 0.05 per cent ascorbic acid.

a. Operations in the first category are differ according to the type of fruit which to be processed.
b. Second category type of operations are similar for all fruit species. Partial elimination of cellulose is achieved with a continuous centrifugal separator; the resulting juice is then processed in order to adjust sugar and acid content for viscosity.

Sugar (about 8-10 per cent) is added as a syrup (in water or in the juice of same fruit obtained by pressure). Acidity is adjusted with citric or tartaric acid. The adjusted juice is then deaerated under vacuum at about 40° C. This step aims at avoiding oxidative reactions and vitamin C loss reduction. An important subsequent step is an intensive homogenisation (under pressure at 150-180 A) in order to obtain particles with dimensions below 100. The homogenised juice obtained is then continuously pasteurized in plate heat exchanger equipment at a temperature of about 130° C, cooled down to about 90° C and aseptically packed in receptacles.

The principal characteristics of fruit "nectars" are uniformity and stability of the content provided by the advanced disintegration of fruits. Stability can be obtained by increasing product viscosity by adding pectin for fruit which is deficient in this component. In order to avoid "separation", intensive homogenisation is carried out as described above. Fruit "nectars" contain all the important components of the original fruit and to a large extent maintain their taste and flavour. The sugar/acidity (as citric acid) ratio is to a large extent determined by the type of fruit and the correction applied; for example, this ratio is 30 for apricots, 40 for peaches, 160 for pears, etc.

MANGO AND GUAVA PROCESSING TECHNOLOGIES

Mango Processing Technologies

Mangoes are processed at two stages of maturity. Green fruit is used to make chutney, pickles, curries and dehydrated products. The green fruit should be freshly picked from the tree. Fruit that is bruised, damaged, or that has prematurely fallen to the ground should not be used. Ripe mangoes are processed as canned and frozen slices, purée, juices, nectar and various dried products. Mangoes are processed into many other products for home use and by cottage industry.The mango processing presents many problems as far as industrialization and market expansion is concerned. The trees are alternate bearing and the fruit has a short storage life; these factors make it difficult to process the crop in a continuous and regular way. The large number of varieties with their various attributes and deficiencies affects the quality and uniformity of processed products.

The lack of simple, reliable methods for determining the stage of maturity of varieties for processing also affects the quality of the finished products. Many of the processed products require peeled or peeled and sliced fruit. The lack of mechanised equipment for the peeling of ripe mangoes is a serious bottleneck for increasing the production of these products.

A significant problem in developing mechanised equipment is the large number of varieties available and their different sizes and shapes. The cost of processed mango products is also too expensive for the general population in the areas where most mangoes are grown. There is, however, a considerable export potential to developed countries but in these countries the processed mango products must compete with established processed fruits of high quality and relatively low cost.

Green Mango Processing

Pickles

The optimum stage of maturity should be determined for each variety used to make pickles. There are two classifications of pickles - salt pickles and oil pickles. They are processed from whole and sliced fruit with and without stones. Salt is used in most pickles. The many kinds of pickles vary mainly in the proportions and kinds of spices used in their preparation. One basic recipe for the study of the preparation and storage of pickles in oil is as follows:

Mango pieces	250 g	Tumeric powder	2 to 4 g
Salt	60 g	Fenugreek seeds	2 to 4 g
Mustard powder	30 g	Bengal gram seeds	2 to 4 g
Chili powder	20 g	Gingelly oil	20 to 30 g

The ingredients are mixed together and filled into wide-mouthed bottles of 0.5 kg capacity. Three days later the contents are thoroughly mixed and refilled into the bottles. Extra oil is added to form a 1-2 cm layer over the pickles.

Chutney

The product is prepared from peeled, sliced or grated unripe or semi-ripe fruit by cooking the shredded fruit with salt over medium heat for 5 to 7 minutes, mixed and then sugar, spices and vinegar are added. Cook over moderate heat until the product resembles a thick purée, add remaining ingredients and simmer another 5 min. Cool and preserve in sterilised jars. Spices usually include cumin seeds, ground cloves, cinnamon, chili powder, ginger and nutmeg. Other ingredients such as dried fruits, onions, garlic and nuts may be added. Drying/ dehydration. immature fruit is peeled and sliced for sun-drying. The dried mango slices can be powdered to make a product called amchoo. The use of blanching, sulphuring and mechanical dehydration gives a product with better colour, nutrition, storability and fewer microbiological problems.

Ripe Mango Processing

Puree

Mangoes are processed into purée for re-manufacturing into products such as nectar, juice, squash, jam, jelly and dehydrated products. The purée can be preserved by chemical means, or frozen, or canned and stored in barrels. This allows a supply of raw materials during the remainder of the year when fresh mangoes are not available. It also provides a more economical means of storage compared with the cost of storing the finished products, except for those which are dehydrated, and provides for more orderly processing during peak availability of fresh mangoes. Mangoes can be processed into purée from whole or peeled fruit. Because of the time and cost of peeling, this step is best avoided but with some varieties it may be necessary to avoid off-flavours which may be present in the skin. The most common way of removing the skin is hand-peeling with knives but this is time-consuming and expensive. Steam and lyepeeling have been accomplished for some varieties.

Several methods have been devised to remove the pulp from the fresh ripe mangoes without hand-peeling. A simplified method is as follows: the whole mangoes were exposed to atmospheric steam for 2 to 2 1/2 min in a loosely covered chamber, then transferred to a stainless steel tank. The steam-softened skins allowed the fruit to be pulped by a power stirrer fitted with a saw-toothed propeller blade mounted 12.7 to 15.2 cm below a regular propeller blade. The pulp is removed from the seeds by a continuous centrifuge designed for use in passion fruit extraction. The pulp material is then passed through a paddle pulper fitted with a 0.084 cm screen to remove fibre and small pieces of pulp. Mango purée can be frozen, canned or stored in barrels for later processing. In all these cases, heating is necessary to preserve the quality of the mango purée. In one process, purée is pumped through a plate heat exchanger and heated to 90°C for 1 min and cooled to 35° C before being filled into 30 lb tins with polyethylene liners and frozen at -23.50 C. In an other process, pulp is acidified to pH 3.5, pasteurized at 90°C, and hot-filled into 6 kg high-density bulk polyethylene containers that have been previously sterilised with boiling water.

The containers are then sealed and cooled in water. This makes it possible to avoid the high cost of cans. Wooden barrels may be used to store mango pulp in the manufacture of jams and squashes. The pulp is acidified with 0.5 to 1.0 per cent citric acid, heated to boiling, cooled, and SO2 is added at a level of 1000 to 1500 ppm in the pulp. The pulp is then filled into barrels for future use.

Slices

Mango slices can be preserved by canning or freezing, and recent studies have shown the feasibility of pasteurized-refrigerated and dehydro-canned slices. The quality of the processed product in all of these procedures will be dependent upon selection of a suitable variety along with good processing procedures. Thermal process canning of mango slices in syrup is the most widely used preservation method.

Beverages

The commercial beverages are juice, nectar and squash. Mango nectar and juice contain mango purée, sugar, water and citric acid in various proportions depending on local taste, government standards of identity, pH control, and fruit composition of the variety used. Mango squash in addition to the above may contain SO_2 or sodium benzoate as a preservative. Other food grade additives such as ascorbic acid, food colouring, or thickeners may be used in mango beverages.

A short description of finished products found in literature is as follows:

- Mango juice: prepared by mixing equal quantities of pulp (purée) and water together and adjusting the total soluble solids (TSS) and acidity to taste (12 to 15 per cent TSS and 0.4 to 0.5 per cent acidity as citric acid);
- Mango nectar containing 25 per cent purée can be prepared using the following procedure.

Table. Brix of Puree

Nectar components	15	17	20
Purée	100	100	100
Sugar	45	43	40
Water	255	257	260

Commercial processing conditions may require the use of a preservative. The pH is adjusted to approximately 3.5 by adding citric acid as a 50 per cent solution. The time of heat processing will vary with filling temperature, can size and viscosity of the juice or nectar. Mango squash may be prepared according to flow-sheet; the finished product may contain 25 per cent juice, 45 per cent TSS and 1.2 to 1.5 per cent acidity and may be preserved with sulphur dioxide (350ppm) or sodium benzoate (1000 ppm) in glass bottles. Mango squash simplified flow-sheet.

Ingredients

Mango pulp	900	900
Sugar	900	1100
Citric acid	18	15
Water	900	900

Mangoes are washed, stored, peeled with stainless steel knives. The pulp is prepared by using a pulper with fine sieve (0.025-in); Sugar is mixed with water and citric acid = syrup; The pulp is added to the syrup and mixed well; The mixture is strained trough cloth; The squash is heated at 85° C and bottles are filled and closed. For additional heat treatment bottles may need to be maintained at a product temperature of 80°C for 30 minutes if the product is to be processed without preservatives. The bottles are then left to cool in water and stored at room temperature. Two negative points must be avoided: presence of air bubbles (which is a source of quick deterioration) and separation of squash solids (giving an undesirable appearance). The means to avoid these two phenomena are described in the fruit juices section.

A type of "squash type" beverage may also be manufactured with '/a pulp + '/a water + i/a sugar and pH adjusted to 3.7 by addition of citric acid. Using different sieve sizes affects the quality and reduces air bubbles to a certain extent but homogenisation and de-aeration of purée or squash seem to be important in order to avoid separation and air bubbles. The squash quality is evaluated on the basis of the following characteristics: pH, titrable acidity, soluble solids, ascorbic acid (by 2,6 dichlorophenol indophenol method), specific gravity.

Dried/Dehydrated

Ripe mangoes are dried in the form of pieces, powders, and flakes. Drying procedures such as sun-drying, tunnel dehydration, vacuum-drying, osmotic dehydration may be used. Packaged and stored properly, dried mango products are stable and nutritious.

One described process involves as pre-treatment dipping mango slices for 18 hr (ratio 1:1) in a solution containing 40° Brix sugar, 3000 ppm SO_2, 0.2 per cent ascorbic acid and 1 per cent citric acid; this method is described as producing the best dehydrated product. Drying is described using an electric cabinet through flow dryer operated at 60° C. The product showed no browning after 1 year of storage. Drum-drying of mango purée is described as an efficient, economical process for producing dried mango powder and flakes. Its major drawback is that the severity of heat preprocessing can produce undesirable cooked flavours and aromas in the dried product. The drum-dried products are also extremely hydroscopic and the use of in-package desiccant is recommended during storage.

Canning

This preservation technology is described in various technological flow-sheets in this bulletin. Mango bar or "fruit leather" is presented in various flow-sheets.

GUAVA PROCESSING TECHNOLOGIES

Guava Purée

Guava purée is used in the manufacture of guava nectar, various juice drink blends and in the preparation of guava jam. The washed sound fruit is first passed through a chopper or slicer to break up the fruit and this material is fed into a pulper. The pulper will remove the seeds and fibrous pieces of tissue and force the reminder of the product through a perforated stainless steel screen. The holes in the screen should be between 0.033 and 0.045 in. The machine should be fed at a constant rate to ensure efficient operation.

The puréed material coming from the pulper is next passed through a finisher. The finisher is equipped with a screen containing holes of approximately 0.020 in. The finisher will remove the stone cells from the fruit and provide the optimum consistency to the product. Perhaps the best way to preserve the guava purée is by freezing and the material passing through the finisher can be packaged and frozen with no further treatment. It is not necessary to heat the product to inactivate enzymes or for other purposes. The material can be frozen in a number of types of cartons and cans; however, a fibre box with a plastic bag inside is commonly used and is probably the less expensive.It is also possible to can and heat process the guava purée and this can be accomplished by heating the purée to 195° F in an open double bottom kettle, filling into cans, closing the cans, inverting the cans for a few seconds, followed by cooling. Cans should be cooled rapidly to approximately 100° F before they are cased and stacked into warehouses.

Guava juice and Concentrate

Guava juice can be used in the manufacture of a clear guava jelly or in various drinks. A clear juice may be prepared from guava purée that is depectinised enzymatically. About 0.1 per cent pectin-degrading enzyme is mixed into the purée at room temperature; heating of the product at approximately 120° F will greatly speed the action of the enzyme. After 1 hr. clear juice is separated from the red pulp by centrifuging or by pressing in a hydraulic juice press. A batch-type or continuous-flow centrifuge can be used on the depectinised purée with no further treatment.The clear juice after centrifuge or after press (and subsequent filtration) can be preserved by freezing or by pasteurization in hermetically sealed cans. For shipment to overseas markets it may be advantageous to concentrate either the purée or the juice.

BANANA AND PLANTAIN PROCESSING TECHNOLOGIES

Traditional Processing

Most of the world's bananas are eaten either raw, in the ripe state, or as a cooked vegetable, and only a very small proportion are processed in order to obtain a storable product. This is true both at a traditional village level with both dessert and cooking bananas and when considering the international trade in dessert bananas. In general, preserved products do not contribute significantly to the diet; however, in some localised areas the products are important in periods when food are scarce. Probably the most widespread and important product is flour preparation from unripe banana and plantains by sun-drying. In Uganda, dried slices known as "mutere" are prepared for storage from green bananas, the dried slices being either used directly for cooking or after grinding into a flour. "Mutere" is used chiefly as a famine reserve and does not feature largely in the diet under normal conditions.

In Gabon, plantains are sometimes made into dried slices which can be stored and used on long journeys, and plantains are used in Cameroon to prepare dried pieces which are stored and ground as needed into flour for use in cooking a paste known as "fufu". Dried green banana slices are also used in parts of South and Central America and West Indies for preparing flour. The other nutritionally important product is beer which is a major product in Uganda, Rwanda and Burundi where green banana utilisation is particularly high.

Preservation Methods and Processes

Drying

Both ripe and unripe bananas and plantains are normally peeled and sliced before drying, although banana figs are sometimes prepared from whole ripe fruit. Sun drying is the most widespread technique where the climate is suitable but drying in ovens or over fires is also practiced. In west Africa, plantains are often soaked and sometimes parboiled before drying. The slices of unripe fruit are normally spread out on bamboo frameworks; or a cemented area; or on a mat; or on a swept-bare patch of earth; or on a roof; or sometimes on stones outcrops or sheets of corrugated iron. Oven-drying of ripe bananas is practiced in Polynesia as a mean of preserving the fruits, which are then wrapped in leaves and bound tightly to store until needed. In East Africa a method has been reported that involves drying the peeled bananas on a frame placed over a fire for 24 hr before drying in the sun, to accelerate the process.

Product Stability and Storage Problems

There is little experimental data on the storage life of the traditionally made banana and plantain products.

Potential for Scaling up of Traditional Processes to Industrial Level

Many banana products are now produced on an industrial scale, including the traditional banana figs and flour, and the processing techniques. One of the main problems encountered has been the susceptibility of banana products to flavour deterioration and discolouration and in the past many products reaching the market have been of poor quality.

A great deal of research has been directed to overcoming these problems, although however good the resultant products are they cannot compare in flavour and other characteristics with the fresh banana fruit. Indeed, an important constraint on the large-scale development of banana processing is the lack of demand for banana products since the fresh fruit is available throughout the year in most parts of the tropical world. The production of beer from banana and plantains has not been scaled up to an industrial level, and while an important product in localised areas of tropical Africa, the market is rapidly declining in favour of European-type brews produced locally.

Industrial Processing

Products and Uses

The main commercial products made from bananas are canned or frozen purée, dried figs, banana powder, flour, flakes, chips (crisps), canned slices and jams. Banana products can be divided roughly into two types - those for direct consumption, such as figs, and those for use in food manufacturing industry, for example purées and powder.

Banana figs, or fingers as they are sometimes known, are usually whole, peeled fruit carefully dried so as to retain their shape, although sometimes the fruit is sliced or halved to facilitate drying. Banana and plantain chips (crisps) are thinly sliced pieces of fruit fried in oil and eaten as a snack like potato chips (crisps). The main use of canned slices is in tropical fruit salads. Banana flakes are used as a flavouring or in breakfast cereals. Banana purée find use mainly in the production of baby foods. Banana flour is said to be highly digestible and is used in baby and invalid foods, but can also be used in the preparation of bread and beverages. Banana powder is used chiefly in the baking industry for the preparation and fillings for cakes and biscuits and is also used for invalid and baby foods.

Processing Technology

In general, to obtain a good-quality product from ripe-bananas the fruit is harvested green and ripened artificially under controlled conditions at the processing factory. After ripening, the banana hands are washed to remove dirt and any spray residues, and peeled. Peeling is almost always done by hand using stainless steel knives, although a mechanical peeler for ripe bananas has been developed, capable of peeling 450 Kg of fruit per hour. The peeling of unripe bananas and plantains is facilitated by immersing the fruit in hot water. For example, immersion in water at 70-75 ° C for 5 min. has been suggested as an aid for peeling green bananas for flour production, while the peeling of green bananas for freezing has been facilitated by immersion in water at 93° C for 30 min.

Banana Figs

Fully ripe fruits with a sugar content of about 19.5 per cent are used and are treated with sulphurous acid after peeling, then dried as soon as possible afterwards. Various drying systems have been described using temperatures between 50 and 82° C for 10 to 24 hr to give a moisture content ranging from 8 to 18 per cent and a yield of dried figs of 12 to 17 per cent of the fresh banana on the stem.

One factory in Australia uses a solar heat collector on the roof to augment the heat used for drying bananas. Bananas can also be dried by osmotic dehydration, using a technique which involves drawing water from 1/4-in. thick banana by placing them in a sugar solution of 67 to 70 deg. Brix for 8 to 10 hr. followed by vacuum-drying at 65 to 70° C, at a vacuum of 10 mm Hg for 5 hr. The moisture content of the final products is 2.5 per cent or less, much lower than that achieved by other methods.

Banana Puree

Banana purée is obtained by pulping peeled, ripe bananas and then preserving the pulp by one of three methods: canning aseptically, acidification followed by normal canning, or quick-freezing. The bulk of the world's purée is processed by the aseptic canning technique. Peeled, ripe fruits are conveyed to a pump which forces them through a plate with 1/4-in. holes, then onto a homogeniser, followed by a centrifugal de-aerator, and into a receiving tank with 29in. vacuum, where the removal of air helps prevent discolouration by oxidation.

The purée is then passed through a series of scraped surface heat exchangers where it is sterilised by steam, partially cooled, and finally brought to filling temperature. The sterilised purée is then packed aseptically into steam-sterilised cans which are closed in a steam atmosphere.

Banana Slices

Several methods for canning of banana slices in syrup are used. Best-quality slices are obtained from fruit at an early stage of ripeness. The slices are processed in a syrup of 25 deg. Brix with pH about 4.2, and in some processes calcium chloride (0.2 per cent) or calcium lactate (0.5 per cent) are added as firming agents. A method for producing an intermediate-moisture banana product for sale in flexible laminate pouches has been developed. Banana slices are blanched and equilibrated in a solution containing glycerol (42.5 per cent), sucrose (14.85 per cent), potassium sorbate (0.45 per cent), and potassium metabisulphite (0.2 per cent) at 90 deg. C for 3 min. to give a moisture content of 30.2 per cent.

Banana Powder

In the manufacture of banana powder, fully ripe banana pulp is converted into a paste by passing through a chopper followed by a colloid mill. A 1 or 2 per cent sodium metabisulphite solution is added to improve the colour of the final product. Spray- or drum-drying may be used, the latter being favoured as all the solids are recovered. A typical spray dryer can produce 70 kg powder per hour to give yields of 8 to 11 per cent of the fresh fruit, while drum-drying gives a final yield of about 13 per cent of the fresh fruit. In the latter method the moisture content is reduced to 8 to 12 per cent and then further decreased to 2 per cent by drying in a tunnel or cabinet dryer at 60° C.

Banana Flour

Production of flour has been carried out by peeling and slicing green fruit, exposure to sulphur dioxide gas, then drying in a counter-current tunnel dryer for 7 to 8 hr. with an inlet temperature of 75° C and outlet temperature of 45° C, to a moisture content of 8 per cent, and finally milling.

Banana Chips (Crisps)

Typically, unripe peeled bananas are thinly sliced, immersed in a sodium or potassium metabisulphite solution, fried in hydrogenated oil at 180 to 200° C, and dusted with salt and an antioxidant. Alternatively, slices may be dried before frying and the antioxidant and salt added with the oil. Similar processes for producing plantain chips have been developed.

Banana Beverages

In a typical process, peeled ripe fruit is cut into pieces, blanched for 2 min. in steam, pulped and pectolytic enzyme added at a concentration of 2 g enzyme per 1 kg pulp, then held at 60 to 65° C and 2.7 to 5.5 pH for 30 min. In a simpler method, lime is used to eliminate the pectin. Calcium oxide (0.5 per cent) is added to the pulp and after standing for 15 min. this is neutralised giving a yield of up to 88 per cent of a clear, attractive juice.

In another process banana pulp is acidified, and steam-blanched in a 28-in Hg vacuum which ensures disintegration and enzyme inactivation. The pulp is then conveyed to a screw press, the resulting purée diluted in the ratio 1:3 with water, and the pH adjusted by further addition of citric acid to 4.2 to 4.3, which yields an attractive drink when this is centrifuged and sweetened.

Jam

A small amount of jam is made commercially by boiling equal quantities of fruit and sugar together with water and lemon juice, lime juice or citric acid, until setting point is reached.

Product Stability and Spoilage Problems

All dried banana products are very hydroscopic and susceptible to flavour deterioration and discolouration, but this can be overcome to some extent by storing in moisture-proof containers and sulphiting the fruit before drying to inactivate the oxidases. The dried products are also liable to attack by insects and moulds if not stored in dry conditions, although disinfestation after drying by heating for 1 hr to 80° C or by fumigation with methyl bromide ensures protection against attack. Banana powder is said to be stored for up to a year commercially and flakes have been stored in vacuum-sealed cans with no deterioration in moisture, colour or flavour for 12 months. Banana chips tend to have a poor storage life and to become soft and rancid. However, chips treated with an antioxidant have been stored satisfactorily at room temperature in hermetically sealed containers up to 6 months with no development of rancidity.

Quality Control Methods

In general a good quality product is obtained if fruit is harvested at the correct stage of maturity and, where appropriate, ripened under controlled conditions. For example, in the case of banana figs, the fruit should be fully mature (sugar content of 19.5 per cent or above) or the final product is liable to be tough and lacking in flavour. However, if over-ripe fruit is used, the figs tend to be sticky and dark in colour, so the fruit must be fully yellow but still firm.

For banana flour, which is prepared from unripe bananas, the fruit is harvested at three-quarters the full-ripe stage and is processed within 24 hr. prior to the onset of ripening. If less mature fruit is used, the flour tastes slightly astringent and bitter due to the tannin content. Bananas harvested between 85 and 95 days after the emergence of the inflorescence, with a pulp-to-peel ratio of about 1.7, were considered to be most suitable for the deep-fat frying.

Other criteria suggested for assessing maturity were beta-carotene and reducing sugar content, both of which increase with increasing maturity and pH which decreases as the fruit ripens, and these should be, respectively, about 2000 μg/100 g, less than 1.5 per cent and 5.8 or above. Browning was found to occur if the sugar content was higher than 1.5 per cent. The determination of crude fat in processed chips is also considered to be a necessary quality control measure. It is important to remove all impurities prior to processing of products, and this is done by washing to remove dirt and spray residues and control on the processing line so that substandard fruit can be removed.

Preparation Methods for Fresh Bananas and Plantains

The main ways of preparing fresh bananas for consumption are boiling or steaming, roasting or baking and frying. Boiling followed by pounding into "fufu" is also widely adopted in certain areas of the tropics.

Boiling or Steaming

Plantains and bananas are often prepared simply by boiling in water, either in their peel or after peeling, and either ripe or unripe; if unripe, the fruit is scraped thoroughly after peeling to remove all traces of fibrous material. The boiled fruit is eaten alone or more usually accompanied by a sauce. This preparation technique is widely used in West Africa.

Roasting or Baking

Unpeeled or peeled fruit, either ripe or unripe, is roasted simply by placing in the ashes of a fire or in an oven. This method is widely used in West Africa, East Africa and the South Pacific islands. For example, ripe plantains are placed unpeeled in an oven and when partly brown and tender, removed and peeled, then replaced in the oven and roasted evenly.

Frying

Ripe or unripe plantains or bananas are often peeled, sliced and cooked in oil, particularly in West Africa and in parts of South America and the West Indies. Similar products are also made in East Africa. Typically, ripe plantains are peeled, cut into slices or split lengthways, and fried in palm oil or with groundnut oil, the pieces being served either hot with a sauce or with fried eggs, or cold as a snack.

Pounding

Pounding is a process, used particularly in West Africa, for preparing most perishable staple food crops including plantains, cassava, yams and cocoyams to obtain a paste or dough known as "fufu" (also spelled "foofoo", "foutou", "foufou"). The plantains are peeled or boiled and peeled after boiling and pounded in a wooden mortar, the resulting paste normally being eaten with soup or a spiced sauce of meat and vegetables, but sometimes after wrapping in leaves and steaming.

7

Soil Quality

ORGANIC MATTER TO REPLENISH THE SOIL

Carrots and onions are prime examples of low-residue crops that leave little organic matter to replenish the soil after harvest. Potatoes produce high amounts of organic matter per acre, but most is removed in the tubers. Radishes and leafy crops like lettuce and spinach have short growing seasons and return little residue even with multi-cropping. To compound the challenge of maintaining SOM, frequent and intense cultivation are generally used in the production of these crops.

Methods discussed in the previous section for increasing SOM—reducing tillage, improving soil structure, and recycling excess plant nutrients—are all suitable for vegetable crops. Application of manures, composts, biosolids, and other organic wastes are good ways to add organic matter. If biosolids are used, they should be low trace-metal, "clean" sludges that can be applied to food crops with no restrictions.

Soil preparation and management operations such as preparing seedbeds, controlling and cultivating weeds, burying crop residue (*e.g.*, for disease control), and building raised beds are important in vegetable production. These operations improve certain soil conditions that otherwise limit crop growth and development. However, each operation also accelerates the loss of SOM by mixing and aerating the soil. And excessive tillage severely degrades soil structure. Loss of structure makes soil prone to compaction and related problems like reduced water infiltration and poor root growth. Tillage operations should be selected carefully, because in some situations growers must weigh short-term benefits to the current crop against long-term reductions in soil quality. Growing cover crops and preceding vegetables with high-residue rotation crops, such as small grains and forages, can increase SOM. In addition, many cover crops suppress weeds; legume cover crops add nitrogen; and cereals and other grasses scavenge residual nutrients and keep them from moving off-site. Buckwheat, sorghum-sudangrass, millets, and annual ryegrass are useful summer cover crops that can follow early-season vegetables or precede fall crops. Rye and millet produce large amounts of vegetative growth, adding potentially large amounts of organic matter. It is important to fertilize rye cover crops or the subsequent cash crop adequately to avoid nitrogen tie-up during decomposition. Common winter cover crops include small grains, such as cereal rye, wheat, and oats. Oats often winter-kill and produce much less dry matter but are a good cover crop for wet soils because they are less likely to delay field operations in the spring. Hairy vetch and red clover are common legume cover crops. They can be grown alone or in grass/legume mixtures.

Many vegetable crops have relatively shallow, sparse root systems but are well fertilized because of their high value. This combination of small root systems and high fertilizer rates may lead to another potential soil-quality problem—the possible movement of excess nutrients to surface water or groundwater. Following a good soil-testing programme, banding fertilizer, splitting nitrogen applications, using fertigation, and avoiding excess irrigation can minimize unwanted nutrient movement.

No-till alternatives have not received as much attention in vegetable crops as in agronomic crops, especially in temperate climates with cold winters. Yet, cover crops are used in successful no-till vegetable systems. Transplanting into killed cover crops (*e.g.*, tomato into hairy vetch) has produced the best results to date. Various practices to overcome the challenges of no-till vegetable production are being tested. For example, seeding or transplanting into living mulches like short life-cycle medics may allow for more timely planting and still offer the benefits of reduced tillage. These and other practices may soon become viable alternatives, especially in northern growing regions.

SOIL MANAGEMENT

Plants need nitrogen, phosphorus, and potassium as well as micronutrients, but getting enough nitrogen, and particularly synchronization so that plants get enough nitrogen at the right time (when plants need it most), is likely the greatest challenge for organic farmers. Crop rotation and green manure ("cover crops") help to provide nitrogen through legumes (more precisely, the *Fabaceae* family) which fix nitrogen from the atmosphere through symbiosis with the bacteria rhizobia. Intercropping, which is sometimes used for insect and disease control, can also increase soil nutrients, but the competition between the legume and the crop can be problematic and wider spacing between crop rows is required. Crop residues can be ploughed back into the soil, and different plants leave different amounts of nitrogen, potentially aiding synchronization. Organic farmers also use animal manure (which must be composted), certain processed fertilizers such as seed meal and various mineral powders such as rock phosphate and greensand, a naturally occurring form of potash which provides potassium. Altogether these methods help to control erosion. In some cases pH may need to be amended. Natural pH amemdments include lime and sulfur, but in the U.S. some synthetically compounds such as iron sulfate, aluminum sulfate, magnesium sulfate, and soluble boron products are allowed in organic farming.

Mixed farms with both livestock and crops can operate as ley farms, whereby the land gathers fertility through growing nitrogen-fixing forage grasses such as white clover or alfalfa and grows cash crops or cereals when fertility is established. Farms without livestock ("stockless") may find it more difficult to maintain fertility, and may rely more on external inputs such as imported manure as well as grain legumes and green manures, although grain legumes may fix limited nitrogen because they are harvested. Horticultural farms growing fruits and vegetables which operate in protected conditions are often even more reliant upon external inputs.

WEED CONTROL

After nutrient supply, weed control is the second priority for farmers. Techniques for controlling weeds include handweeding, mulch, corn gluten meal, a natural pre-emergence herbicide, flame, garlic and clove oil, borax, pelargonic acid, table salt, solarization (which involves spreading clear plastic across the ground in hot weather for 4–6 weeks), vinegar, and various other homemade remedies. One recent innovation in rice farming is to introduce ducks and fish to wet paddy fields, which eat both weeds and insects.

Controlling other Organisms

Organisms aside from weeds which cause problems include arthropods and nematodes. Fungi and bacteria can cause disease. Insect pests are a common problem, and insecticides, both non-organic and organic, are controversial due to their environmental and health effects. One way to manage insects is to ignore them and focus on plant health, since plants can survive the loss of about a third of leaf area before suffering severe growth consequences. To avoid using insecticides, one can select naturally-resistant plants, put bags around the plants, remove dying material such as leaves, fruit, and diseased plants, covering plants with a solid barrier ("row cover"), hosing, encouraging and releasing beneficial organisms and beneficial insects, planting companion plants and polycultures, various traps, sticky cards (which can also be used to assess insect prevalence), and season extension.

Biological pest control uses natural predators to control pests. Recommended beneficial insects include minute pirate bugs, big-eyed bugs, and to a lesser extent ladybugs (which tend to fly away), all of which eat a wide range of pests. Lacewings are also effective, but tend to fly away. Praying mantis tend to move slower and eat less heavily. Parasitoid wasps tend to be effective for their selected prey, but like all small insects can be less effective outdoors because the wind controls their movement. Predatory mites are effective for controlling mites. Several of pesticides approved for organic use have been called green pesticides such as spinosad and neem. Generally, but not necessarily, organic pesticides are safer and more environmentally friendly than synthetic pesticides. The main three organic insecticides used are Bt (a bacterial toxin), pyrethrum and rotenone. Surveys have found that fewer than 10 per cent of organic farmers use these pesticides regularly; one survey found that only 5.3 per cent of vegetable growers in California use rotenone while 1.7 per cent use pyrethrum. As of 2005, the controversial and highly toxic insecticide rotenone was theoretically approved for U.S. organic farmers, but no products had been reviewed by the Organic Materials Review Institute. Nicotine sulfate may also be used; although it breaks down quickly, it is extremely toxic, nearly as toxic as aldicarb. Less toxic but still effective organic insecticides include neem, spinosad, soaps, garlic, citrus oil, capsaicin (repellent), *Bacillus popillae*, *Beauvaria bassiana*, and boric acid. Pesticides should be rotated to minimize pest resistance.

The first disease control strategy involves keeping the area clean by removing diseased and dying plants and ensure that the plants are healthy by maintaining water and fertilization. Compost tea is sometimes promoted and can be effective, but there is concern over whether these are ineffective or even harmful. Polycultures reduce the ability of disease to spread. Disease-resistant cultivars can be purchased. Organic fungicide include the bacteria Bacillus subtilis, Bacillus pumilus, and Trichoderma harzianum which are mainly effective for diseases affecting roots. Bordeaux mix contains copper, which can be used as an organic fungicide in various forms. Sulfur is effective against fungus as well as some insects. Lime sulfur is also available, but can damage plants. Potassium and sodium bicarbonate are also effective against fungus. Some plant activators, which increase plants' defence systems, are considered organic although most are synthetic. Other synthetic fungicides not generally allowed are classified as protectants and systemics.

Standards

Standards regulate production methods and in some cases final output for organic agriculture. Standards may be voluntary or legislated. As early as the 1970s organic producers could be voluntarily certified by private associations. In the 1980s, governments began to produce organic production guidelines. Beginning in the 1990s, a trend towards legislation of standards began, most notably with the 1991 EU-Eco-regulation developed for European Union, which set standards for 12 countries, and a 1993 UK programme. The EU's programme was followed by a Japan programme in 2001, and in 2002 the United States created the National Organic Programme (NOP). As of 2007 over 60 countries have regulations on organic farming. In 2005 IFOAM created the Principles of Organic Agriculture, an international guideline for certification criteria. Typically the agencies do not certify individual farms, but rather accredit certification groups. Materials used in organic production and foods are tested independently by the Organic Materials Review Institute.

Composting

Under USDA organic standards, manure must be subjected to proper thermophilic composting and allowed to reach a sterilizing temperature. If raw animal manure is used, 120 days must pass before the crop is harvested.

Economics

The economics of organic farming, a subfield of agricultural economics, encompasses the entire process and effects of organic farming in terms of human society, including social costs, opportunity costs, unintended consequences, information asymmetries, and economies of scale. Although the scope of economics is broad, agricultural economics tends to focus on maximizing yields and efficiency at the farm level. Mainstream economics takes an anthropocentric approach to the value of the natural world: biodiversity, for example, is considered beneficial only to the extent that it is valued by people and increases profits. Some governments such as the European Union subsidize organic farming, in large part because these countries believe in the external benefits of reduced water use, reduced water contamination by pesticides and nutrients of organic farming, reduced soil erosion, reduced carbon emissions, increased biodiversity, and assorted other benefits. Organic farming is labour and knowledge-intensive whereas conventional farming is capital-intensive, requiring more energy and manufactured inputs. Organic farmers in California have cited marketing as their greatest obstacle.

Geographic Producer Distribution

The markets for organic products are strongest in North America and Europe, which as of 2001 are estimated to have $6 and $8 billion respectively of the $20 billion market. However, as of 2007 organic farmland is distributed across the globe. Australasia has 39 per cent of the total organic farmland with Australia's 11.8 million hectares, but 97 percent of this land is sprawling rangeland, which results in total sales of approximately 5 per cent of US sales. Europe has 23 percent of total organic farmland (6.9 million hectares), followed by Latin America with 19 percent (5.8 million hectares). Asia has 9.5 percent while North America has 7.2 percent. Africa has a mere 3 percent. Besides Australia, the countries with the most organic area are Argentina (3.1 million hectares), China (2.3 million hectares), and the United States (1.6 million hectares). Much of Argentina's organic farmland is pasture, like that of Australia. Italy, Spain, Germany, Brazil, Uruguay, and the UK follow the United States by the amount of land managed organically.

Growth

As of 2001, the estimated total market value of certified organic products was estimated to be $20 billion. By 2002 this was $23 billion and by 2007 more than $46 billion according to Organic Monitor. In recent years both Europe and North America have experienced strong growth in organic farmland. However, this growth has occurred under different conditions. While the European Union has shifted agricultural subsidies to organic farmers in recognition of its environmental benefits, the United States has taken a free market approach. As a result, as of 2007 4 percent of the European Union's farmland was organically managed compared to just 0.6 percent of United States farmland. IFOAM's most recent edition of The World of Organic Agriculture: Statistics and Emerging Trends 2009 lists the countries which had the most hectares in 2007. The country with the most organic land is Australia with more than 12 million hectares, followed by Argentina, Brasil and the US. In total 32.2 million hectares were under organic management in 2007. For 1999 11 million hectares of organically managed land are reported. In recent years organic agriculture has grown tremendously. Considering this rapid growth, it is within the nature of organic farming to keep it from becoming a large scale industrial business as conventional farming has become. Duram, Leslie. Good Growing. Santa Cruz: Bison Books, 2005.

Productivity and Profitability

A 2006 study suggests that converted organic farms have lower pre-harvest yields than their conventional counterparts in developed countries (92 per cent) and that organic farms have higher pre-harvest yields than their low-intensity counterparts in developing countries (132 per cent). The researcher attributes this to a relative lack of expensive fertilizers and pesticides in the developing world compared to the intensive, subsidy-driven farming of the developed world. Nonetheless, the researcher purposely avoids making the claim that organic methods routinely outperform green-revolution (conventional) methods.

This study incorporated a 1990 review of 205 crop comparisons which found that organic crops had 91 per cent of conventional yields. A major US survey published in 2001, analyzed results from 150 growing seasons for various crops and concluded that organic yields were 95-100 per cent of conventional yields. Lotter reports that repeated studies have found that organic farms withstand severe weather conditions better than conventional farms, sometimes yielding 70-90 per cent more than conventional farms during droughts. A 22-year farm trial study by Cornell University published in 2005 concluded that organic farming produces the same corn and soybean yields as conventional methods over the long-term averages, but consumed less energy and used zero pesticides. The results were attributed to lower yields in general but higher yields during drought years. A study of 1,804 organic farms in Central America hit by Hurricane Mitch in 1998 found that the organic farms sustained the damage much better, retaining 20 to 40 per cent more topsoil and smaller economic losses at highly significant levels than their neighbours.

On the other hand, a prominent 21-year Swiss study found an average of 20 per cent lower organic yields over conventional, along with 50 per cent lower expenditure on fertilizer and energy, and 97 per cent less pesticides. A long-term study by U.S Department of Agriculture Agricultural Research Service (ARS) scientists concluded that, contrary to widespread belief, organic farming can build up soil organic matter better than conventional no-till farming, which suggests long-term yield benefits from organic farming. An 18-year study of organic methods on nutrient-depleted soil concluded that conventional methods were superior for soil fertility and yield in a cold-temperate climate, arguing that much of the benefits from organic farming are derived from imported materials which could not be regarded as "self-sustaining".

While organic farms have lower yields, organic methods require no synthetic fertilizer and pesticides. The decreased cost on those inputs, along with the premiums which consumers pay for organic produce, create higher profits for organic farmers. Organic farms have been consistently found to be as or more profitable than conventional farms with premiums included, but without premiums profitability is mixed. Welsh (1999) reports that organic farmers are more profitable in the drier states of the United States, likely due to their superior drought performance. In 2008 the UN Environmental Programme (UNEP) and UN Conference on Trade and Development (UNCTAD) issued a report which stated that "organic agriculture can be more conducive to food security in Africa than most conventional production systems, and that it is more likely to be sustainable in the long-term". The report assessed 114 projects in 24 African countries, finding that "yields had more than doubled where organic, or near-organic practices had been used" and that soil fertility and drought resistance improved. In 2009, a review concluded that organic production was more profitable in Wisconsin, when including price premiums.

Macroeconomic Impact

Organic methods often require more labour, providing rural jobs but increasing costs to urban consumers.

Motivations

Agriculture in general imposes external costs upon society through pesticides, nutrient run-off, excessive water usage, and assorted other problems. As organic methods minimize some of these factors, organic farming is believed to impose fewer external costs upon society. A 2000 assessment of agriculture in the UK determined total external costs for 1996 of 2343 million British pounds or 208 pounds per hectare. A 2005 analysis of these costs in the USA concluded that cropland imposes approximately 5 to 16 billion dollars ($30 to $96 per hectare), while livestock production imposes 714 million dollars. Both studies concluded that more should be done to internalize external costs, and neither included subsidies in their analysis, but noted that subsidies also influence the cost of agriculture to society. Both focused on purely fiscal impacts. The 2000 review included reported pesticide poisonings but did not include speculative chronic effects of pesticides, and the 2004 review relied on a 1992 estimate of the total impact of pesticides.

Pesticides

Most organic farms use fewer pesticides than conventional farms, some pesticides damage the environment or with direct exposure human health. The main five pesticides used in organic farming are Bt (a bacterial toxin), pyrethrum, rotenone, copper and sulphur. Surveys have found that fewer than 10 per cent of organic farmers use these pesticides regularly; one survey found that only 5.3 per cent of vegetable growers in California use rotenone while 1.7 per cent use pyrethrum.

Reduction and elimination of chemical pesticide use is technologically challenging. Few organic farms manage to eliminate the use of pesticides entirely; organic pesticides are often used to complement other pest control strategies. Pesticide run-off is one of the most significant effects of pesticide use. The USDA Natural Resources Conservation Service tracks the environmental risk posed by pesticide water contamination from farms, and its conclusion has been that "the Nation's pesticide policies during the last twenty six years have succeeded in reducing overall environmental risk, in spite of slight increases in area planted and weight of pesticides applied.

Nevertheless, there are still areas of the country where there is no evidence of progress, and areas where risk levels for protection of drinking water, fish, algae and crustaceans remain high". Pest resistant genetically modified crops have been proposed as an alternative to pesticide use, however concerns over the safety and the long term benefits of genetically modified food, result in the genetic modification being widely opposed in the organic farming movement.

EFFECTS ON SOIL EROSION

Crop rotation can greatly affect the amount of soil lost from erosion by water. In areas that are highly susceptible to erosion, farm management practices such as zero and reduced tillage can be supplemented with specific crop rotation methods to reduce raindrop impact, sediment detachment, sediment transport, surface run-off, and soil loss. Protection against soil loss is maximized with rotation methods that leave the greatest mass of crop stubble (plant residue left after harvest) on top of the soil. Stubble cover in contact with the soil minimizes erosion from water by reducing overland flow velocity, stream power, and thus the ability of the water to detach and transport sediment Soil Erosion and Cill prevent the disruption and detachment of soil aggregates that cause macrospores to block, infiltration to decline, and run-off to increase. This significantly improves the resilience of soils when subjected to periods of erosion and stress.

The effect of crop rotation on erosion control varies by climate. In regions under relatively consistent climate conditions, where annual rainfall and temperature levels are assumed, rigid crop rotations can produce sufficient plant growth and soil cover. In regions where climate conditions are less predictable, and unexpected periods of rain and drought may occur, a more flexible approach for soil cover by crop rotation is necessary. An opportunity cropping system promotes adequate soil cover under these erratic climate conditions. In an opportunity cropping system, crops are grown when soil water is adequate and there is a reliable sowing window. This form of cropping system is likely to produce better soil cover than a rigid crop rotation because crops are only sown under optimal conditions, whereas rigid systems are sown in the best conditions available.

Crop rotations also affect the timing and length of when a field is subject to fallow. This is very important because depending on a particular region's climate, a field could be the most vulnerable to erosion when it is under fallow. Efficient fallow management is an essential part of reducing erosion in a crop rotation system. Zero tillage is a fundamental management practice that promotes crop stubble retention under longer unplanned fallows when crops cannot be planted. Such management practices that succeed in retaining suitable soil cover in areas under fallow will ultimately reduce soil loss

CARBON DIOXIDE FERTILIZATION

Widely recognised as the matriarch of the greenhouse industry, the Dutch industry spreads over some 12000 hectares. Approximately 7000 hectares are to be found in the region known as the Westlands (mostly vegetables), with a further 4000 hectares in the Aalsmeer district (mostly floriculture) and the remainder scattered around the country, with a concentration in the areas around Venlo and Emmen.

The greenhouse industry has been part of the Dutch scenery for decades. For example, in the area of Naaldwijk, in the Westlands, glasshouses currently cover 25 per cent (or 3000 hectares) of the land area. Subsequently glasshouses are hand-in-hand with windmills and canals in representing the lifestyle of The Netherlands. The estimated minimum viable operation is 4000m2, though most new developments range between three and 30 hectares in scale. England is very much a small-scale version of the Dutch industry. Greenhouse production covers approximately 400 hectares and is found throughout the mid to southern regions.

A similar situation is found in Sweden, though, the industry in Sweden is of much lesser significance than the British industry. In Sweden, the technology and practices are a near carbon copy of the Dutch industry, or more accurately, the Dutch industry of 10-20 years ago. The most striking feature of these industries (Dutch and pseudo-Dutch) is the similarity between one operation and the next. These industries demonstrate a base level of technicality, which every operation has attained.

In fact, the difference in technology at the top compared with the bottom of the industry is extremely narrow. In contrast, there are two divergent industries in Spain, similar to the situation found in Australia. The larger, traditional Spanish industry has evolved its own unique greenhouse structure, which has minimal environmental control and is dominantly used for winter production. The smaller, newer industry on the other hand, reflects the world's best standards in technology and greenhouse design. A heavy Dutch and British influence is detectable, with many modern operations, in fact, involving direct international - often Dutch - investment. In total, the greenhouse industry in Spain covers as much as 40000 hectares, though this may be a conservative estimate. The newer industry possibly accounts for no more than two percent of the total industry at present.

The industry in Canada (in the two provinces I was able to visit) more closely spotlights the potential of the Australian industry than anywhere else I visited. A strong Dutch influence is noticeable though there remains a distinct range of technological levels throughout the industry unlike the Dutch industry that varies little between growers. Two key areas of the Ontario industry are the Leamington and the Niagara districts, to the south and southeast respectively of Toronto. The Leamington area is primarily involved in the growing of vegetables while the Niagara peninsula is the largest floricultural production area in Canada.

There are approximately 800-1000 hectares of greenhouses in Ontario, with around half the area used for vegetable production, half for floriculture. Tomato is the single biggest greenhouse vegetable crop in Ontario, with over fifty percent of the harvest exported to the USA. On the West Coast of Canada, the British Columbian greenhouse vegetable industry is around 300 hectares and floriculture accounts for a further 100 hectares of structures. Vegetable and floricultural operations are mixed throughout the region of Vancouver from the Delta region east to Abbotsford. A small proportion of the industry is located on Vancouver Island. A range of greenhouses are used, including glass and plastic clad structures and the level of technology varies significantly from one end of the industry to the other, though this difference does not necessarily correlate with the profitability of an operation. The average size of floricultural greenhouse operations is approximately 1 hectare while vegetable operations are twice this area.

ORGANIC VEGETABLE PRODUCTION INTEGRATED DEVELOPMENT

Organic farming is estimated to be growing at 30 per cent a year worldwide in response to market forces (Troeth, 2001). Export market demand for certified organic produce, especially vegetables, currently exceeds supply (Kinnear, 2000) and in many cases produce attracts premium prices (Fielke, 2001). The Australian organic industry is small relative to the conventional food production industry. This trend is reflected in the Tasmanian vegetable industry where only five, out of approximately 1500 commercial vegetable growers are certified organic. Only one organic vegetable grower is producing on a large-scale commercial basis.

Tasmanian packing and exporting companies have increased their interest in organic produce in response to the global trend for increased consumption. The opportunities are considered substantial enough by some to initiate, develop and support a commercial organic vegetable industry in Tasmania. Export markets have considerable appeal due to demand for year round supply of fresh organic produce and an average price premium of 23 per cent above conventional product. A scarcity of data on the commercial production of intensively cropped organic vegetables, combined with market potential, were the primary drivers of this four-year project. The result was a partnership between the Department of Primary Industries, Water and Environment (DPIWE) and Field Fresh Tasmania (FFT), with funding from the Rural Industries Research and Development Corporation (RIRDC).

Aims/Objectives:

- Test and evaluate, on a commercial scale, organic production protocols for a range of vegetable crops.
- Document case studies assessing the production protocols, cost, net yields, market premiums and cost effectiveness of intensive organic vegetable production.
- Provide information and training in organic production protocols for both existing and prospective organic vegetable growers.

A 10-hectare site, located amidst intensive vegetable production enterprises on the northwest coast of Tasmania, was selected for the trial. Conversion of the site to a certified organic status was conducted during the project and was a principle feature of the work. The trial site consisted of three paddocks, which were split lengthwise to form six units of approximately 1.2 hectares each. A five-year cropping rotation was then overlaid onto the site plan representing an intensive vegetable production system of four years crop and one-year pasture/rest phase. Crops chosen consisted of those considered relatively easy to grow organically (broad beans, snow peas, carrots) as well as those with the highest market demand but greater difficulty to produce organically (onions, shallots).

Although an initial 5-year rotation was selected as a guide at the beginning of the trial, this was reviewed prior to each season and adjustments made according to the market information and the previous season's production data. Production Protocols: Production protocols for the major crops (onions and carrots) and for test crops (shallots, broad beans and snow peas) were progressively developed, adapted and refined throughout the trial for improved performance.

Carrots and Onions: Inter-row spacing configurations were manipulated for greater soil cover and to improve accessibility and effectiveness of weeding machinery. Planting times for carrots were altered to reduce insect burden and to meet market requirements. Addition of organic fertilizer and nitrogen budgeting were incorporated to improve nutritional health of crops. Bed preparation and pre-emergence weeding (flame) were adjusted for weed management purposes in onions. Shallots, broad beans and snow peas: A protocol for shallots was established however, bulbs from the first crop were mostly unmarketable due to a late planting date and a colder than usual season. A second shallot crop was planted in the subsequent season using the former season's protocol.

Planting layout and density of broad beans were manipulated for improved weed control, airflow and accessibility. Snow peas were deemed uneconomical due to high hand harvesting costs and low yields, and were ploughed in as green manures. Yield data from the trial was collated and details are presented in the report. Yields were mostly expressed in terms of net yield (t/ha) of Class I product and pack out (percentage yield of Class I product). The exception is where either low total yield or low product quality occurred, in which instance only gross values are provided.

Carrots and Onions: With the exception of the 2002 crop yields, the packout values for the harvested carrot crops were similar or better than average conventional yields indicating an overall high quality. The carrot crop yield in 2002 was low, primarily due to wet weather and wet soil preventing pre-emergence flame weeding. The yields of organic onions were low compared to conventional yields. A heavy weed burden may have impacted on yield. Cutworm and accidental removal of crop plants with mechanical weeding implements may have also had a minor effect. A change in planting configuration in 2002 contributed to a further decline in yield during that season.

Shallots, Broad Beans and Snow Peas: The quality of product harvested from shallot crops was considered to be of a suitable market quality and comparable with conventional crops. However, due to agronomic difficulties experienced, it is difficult to draw any meaningful conclusions or views relating to yields and associated factors. Difficulties experienced related to a later than ideal planting time and colder growing conditions which were incompatible with variety requirements. Harvested seed from the first broad bean crop was retained for subsequent use. Sufficient product from the second broad bean crop was harvested specifically for test marketing purposes and as a consequence extensive field data on yields and associated factors was not collated. Snowpeas were produced for one season only. Due to the high costs associated with the harvest and packing of crop, the volume harvested was extremely small and not sufficient to produce reliable data for analysis.

Marketing and Economics: Market opportunities were assessed prior to crop selection and planting. Produce from the trial was sent internationally to the United Kingdom, Japan and interstate.

Carrots and Onions: Initial market feedback on the first year's production of carrots was very positive regarding quality, however all the markets indicated that premiums for pre-conversion and in-conversion organic produce are difficult to achieve. Data from the first crop indicated that organic carrot production could cost $100-$130 a net tonne more than conventional carrot crop production. This poses a difficult financial dilemma for the pre-conversion and conversion periods in the absence of a suitable market premium. Despite low onion yields, pack-out figures were high indicating excellent onion quality. The majority of the harvested bulbs fitted market specifications and were exported to the United Kingdom. The bulbs were solid with tight skins presenting them as ideal for long term storage and shipping. Based on this experience they are considered comparable with onions grown using conventional production practices.

Shallots, Broad Beans and Snow Peas: Shallot bulbs from the first crop were mostly unmarketable due to late planting date and colder than normal growing season. In 2002, FFT planted a shallot crop using the protocols established. Problems occurred when irrigation and rain delayed weeding.

Due to the small recovery of Class 1 product in the second season the view was taken that the volume was too small to justify export. Instead, the product was marketed in Australia – at higher prices than those achievable in export markets. Commercial samples of broad beans were airfreighted to Japan and the quality was reported to be of a suitable standard. Difficulties were experienced due to the extremely narrow production window, which prevented a sustainable marketing campaign.

Evaluation was undertaken in Japan on the options of supplying product in complete pod form and individual beans, but both options proved to be commercially non-viable.

Test marketing in Australia was undertaken and although initial quality feedback was positive, fungal deterioration became apparent with some product losses experienced by clients. Demand for broad beans over the production period was limited and clients reported that product demand was strongest during winter months. Whilst the product was considered to have niche interest, the prices were well below the levels required to make it commercially viable. Snow peas were grown in 2001; the crop was sown in two plantings for sequential harvest. The hand harvesting costs for the first planting outweighed the market return. The variety was determined to be unsuitable and achievable prices were extremely low. Although product quality was considered suitable, hand harvest meant production was uneconomic.

Weed and Pest Management: Weed control in the absence of herbicides was the dominant pest management issue in this project. Weed management was particularly challenging in the allium crops as they do not form a crop canopy to assist in competition against weeds. These crops require constant weed control through to harvest.

However, non-herbicide weed management protocols for onion, carrot and broad bean crops were established and are detailed in this report. They consist of a range of well-timed mechanical techniques. Evidence of disease was detected in a number of crops grown at the organic trial site. However, whilst crop diseases occurred at the site, they were not considered a major management issue. Copper and sulphate applications were used as the only form of fungicide at the site; the mixture is preventative and requires regular applications. Overall, the effectiveness of the fungicide treatments was good. Poor results were generally viewed as due to poor timing in application. Overall insect pests, like disease, were not considered a major problem at the organic trial site. However, they were detected in some crops at levels high enough to cause economic damage. Insect pest management included the use of natural pyrethrins.

Agronomic Challenges for Converting to Intensive Organic Vegetable Production: The three principal agronomic challenges identified by the trial work are:

- *Weed Management:* Weeds proved to be the greatest pest management issue associated with intensive organic vegetable production and were a significant factor in the final production costs.
- This project indicated the major challenge to improve the economic viability of intensive organic vegetable production was to decrease the level of hand weeding required.
- *Soil Health and Nutrition:* Soil structural damage was noted at the site and was primarily attributed to heavy vehicle use on moist soil at harvest. In addition, despite legume rotations, available nitrogen decreased rapidly over the trial period. Improvements in soil organic carbon were identified as essential, while pH levels were considered satisfactory. Long term soil monitoring at the site would be beneficial.
- *Skills:* An organic producer needs to be highly skilled with additional and/or different skill sets to conventional farmers. The project team believe that many conventional farmers that may decide to convert would need additional training and skills support particularly through the conversion period. Such support is likely to enhance the success rate of the conversion process and subsequent organic cropping operation.

CHARACTERISTICS AFFECTING VEGETABLE PRODUCTION

The unique combination of climate, landscape, parent material, and living organisms in southwest Florida has greatly influenced the formation of soils in this area. With time, many of these soils have formed distinguishing characteristics, termed diagnostic horizons. These horizons are identified by observing the soil with depth, such as the side of a pit or a road cut. In turn, these diagnostic horizons are useful for classifying soils and relating their characteristics to commercial crop production.

The surface horizon, designated by a capital A, usually appears gray to black and is almost always of sandy texture. Moving down the soil profile and just beneath the A horizon, a leached zone called the E horizon is often found. Nutrients and fine particles including organic matter are moved by water from this horizon. The E horizon is usually much lighter in colour than the surface horizon, often gray to white. Located beneath this leached zone, some soils have a distinct brown or black horizon, called the Bh or spodic horizon. This horizon is composed of organic matter that is leached down the profile and by both physical and chemical means has been deposited in the lower part of the soil profile. Often high in aluminum and iron and usually with a low pH, the Bh horizon is almost always sandy in texture. This horizon impedes water flowing vertically through the soil and causes water to accumulate above this horizon. This water accumulation is referred to as a perched water table and is quite beneficial for maintaining a constant water table for row crop subsurface irrigation. In some soils, the Bt is another diagnostic horizon that may be found below the leaching zone or occasionally with the Bh horizon. This horizon is created by the deposition of clay particles and is usually mottled gray in colour and sandy or sandy loam in texture. Similar to the Bh horizon, the Bt horizon may also allow the formation of a perched water table.

Below these diagnostic horizons, substratum may be present. The C horizon denotes substratum that can be a variety of textures and colours but is usually unconsolidated materials. The R horizon denotes substratum that is limestone bedrock (not shown). Using these diagnostic horizons, most of the mineral soils in southwest Florida can be classified. The USDA-NRCS (Natural Resource Conservation Service) has prepared soil survey reports for each county in Florida. These documents are invaluable for understanding soils used for commercial vegetable operations and contain maps showing the spatial distribution of soils in the landscape. While some soils may be better suited than others for vegetable production, soils often occur in associations and complexes, which describe a mixture of soils within the named mapping unit in the surveys. Thus, soil survey maps are considered accurate but not precise, meaning that the general management, as well as the chemical and physical characteristics within a mapping unit, is accurately described, but any one spot within the landscape may not be precisely described.

Further complicating precise characterization of any specific landscape point is the fact that some of the soil profile layers may be mixed during the construction of drainage ditches and beds. Depending upon the construction of field drainage and vegetable bedding operations, soil material used to create the bed or excavated to form the drainage ditch may actually contain portions of the lower horizons. Thus, vegetables may be planted (transplanted) in portions of the subsoil, rather than in the A horizon where most of the nutrients and organic matter of the original soil are located.

SIGNIFICANCE OF VEGETATIVE PROPAGATION

Rooting in stem cuttings can be important means of vegetative propagation for afforestation purposes. In arid zones, quick establishment of plants with ample root systems is a necessity. In arid regions, water in the form of precipitation is available only in the rainy season, and the plants must be established in suitable conditions of soil moisture. Therefore, rooted stem cuttings are more useful than seed sowing because rooted cuttings are far better able to survive in the stressful environment of the desert than delicate seedlings.

There are various methods of multiplication of mulberry plants. In mulberry species, the stem cuttings readily form roots. Both grafting and layering need time for establishment. Hence, propagation of mulberry through stem cuttings is preferred. Thimann and Behnke-Rogers showed that the rooting of cuttings of many tree species is stimulated by synthetic growth substances. Bose has developed easier and better methods of vegetative propagation by the use of growth substances for ornamental and fruit plants. Bose and Mukherjee used some growth substances to improve rooting in cuttings of *Legerstroemia indica*. Prasad and Dikshit obtained maximum success in rooting with cuttings of essential oil–producing plants treated with growth regulators. Teaotia and Pandey obtained better results in rooting guava stem cuttings with the assistance of growth substances.

FACTORS AFFECTING VEGETATIVE PROPAGATION

More than 50 per cent of the land surface of the developing countries is located in the arid and semiarid zones. In many of these countries, in which more than 80 per cent of the population lives with agricultural and animal husbandry, a tragic and dangerous imbalance is developing between requirements for and available supply of food, fodder, and fuel. Dwindling vegetation cover will adversely affect all facets of rural life in which trees and shrubs generally serve not only as fuel but also as shade and shelter for man, animal, and crops. In the long term, depletion of the natural vegetation will increase ecological fragility and contribute to gradual degradation of the resource base as well as the natural resources themselves.

A practice common among peasants is migration of cattle to neighbouring states or within the state wherever fodder is available. This large-scale migration does immense harm to the delicate ecosystem. Animals usually strip all of the plants from the area; this causes poor regeneration and increased soil erosion, and more areas become barren. This necessitates the utilization of saline wastelands for fodder production as crop cultivation is impossible because of the high salt content of the soil. Enumeration of indigenous saline species showed that very few plants are palatable and their growth pattern is not at an acceptable level for fodder production.

Many taxa of the family Chenopodiaceae are indigenous to arid and saline regions of the world. Their ecological amplitude is very high, and various adaptive features at different levels of the plant life cycle are observed. Many are shrubs, and they offer a tremendous potential for human benefit in making the arid and semiarid lands of the world more productive and useful. To revegetate the salt-affected soils and secondary salinized soils, plants that can survive in arid and saline conditions are needed. Shrubby halophytes of the genus *Atriplex* are particularly adapted to such conditions. The genus *Atriplex* includes several haloxeric fodder species very useful in arid zones. The primary driving force of all animals is the need to finding the right kind of food and enough of it. Food is the burning question in animal society, and the whole structure and activities of the community are dependent upon questions of food supply. Saline and sodic soils are problems of individual localities, and their formation and causes of development must be considered before these soils are put to any economic use. Salt-tolerant plants have been used as forage in arid saline areas for millennia.

The recognition of the value of certain salt-tolerant shrub grass species is reflected in their incorporation in pasture improvement programmes in many salt-affected regions throughout the world. However, reproduction, survival, and multiplication under the inhospitable conditions of arid saline areas are basic needs for any halophytic or glycophytic species. In many halophytes, germination of seeds is usually retarded by high concentrations of salt in the soil. Germination is the most important stage in the life cycle of any species growing in an arid saline environment. Seed germination in saline environments occurs mostly with high precipitation, when soil salinity levels are usually reduced.

It is also known that when seeds are sown in a saline environment, there is a decrease in the rate of germination, delaying completion of germination; moreover, there is a water potential below which germination does not occur. In general, it is agreed that salinity affects germination by creating sufficiently low to inhibit water uptake (osmotic effect) and/or by providing conditions for the entry of ions that may be toxic to the embryo. These constraints affect the different stages of seed germination and seed establishment to varying degrees.

Reduction of germination occurs when halophytes are subjected to salinities above 1 per cent NaCl; increasing salt concentrations also delay germination. Salinity or sodicity and water stress are the most important factors responsible for limiting seed germination and plant growth. To overcome the present environmental stress of saline areas, plants produce a variety of ecological adaptations. Propagation through vegetative means has been used as a method of multiplication for a number of plant species under arid saline conditions.

Among factors affecting rooting of cuttings, the position of the shoot plays an important role. It is reported that without auxin treatment and without leaves, no roots were obtained in cutting of red *Hibiscus* and *Allamenda cathartica*. Vegetative reproduction substitutes for or at least contributes to the reproductive potential of many plants. This statement is more applicable to various halophytic species that are restricted to narrow ecological limits, either in the production of disseminules or by their germination.

Self-layering species of *Atriplex* are at an advantage in establishing themselves in salt-affected soil, which they accomplish faster than other species: the growth of developing roots results in rapid penetration through the upper salty soil layers. Furthermore, roots developing at different nodes are not dependent on a direct supply of water from the soil. Being well supplied with water by the parent plant, roots can penetrate layers of extreme salinity.

VEGETATIVE PROPAGATION IN SALINE PLANTS

The distribution of salinity varies spatially, temporally, qualitatively, and quantitatively. In addition, the responses of plants to salt stress vary during the life cycle of the individual. Phenotype plasticity involving both morphological and physiological changes in response to episodic events is an important characteristic associated with the survival of long-lived plants under highly stressful environmental conditions. Transient reductions in yield in response to salinity may be the result of the adaptive reconstruction of growth habits of a plant. The heterogeneity of saline habitats leads to considerable genetic differentiation among populations as a result of natural selection: an all-purpose genotype capable of growing in a wide range of saline habitats probably does not exist. The growth and productivity of *Atriplex* under conditions of low and erratic rainfall are exceptional, and the adaptation of this species to high salinity makes its introduction very suitable. Agronomic testing, feeding trials, and development of the best agronomic practices are necessary in the evaluation of suitable species for introduction and mass propagation. Normal vegetation, except for some halophytes, cannot survive on saline and sodic soils. Thus, areas having soils of these types are of limited agricultural use unless the salinity is quite mild. Increased salinity has rendered many lands unfit for cultivation. Plant species that are capable of accumulating large quantities of sodium in their tissues are the least sensitive to the presence of salt in the soil. The tolerance of a species to high amounts of absorbed or exchangeable sodium is modified by the pH of the soil and by the accumulation of CO_2. With increasing human and animal populations and the need for greater crop and fodder production, Non-productive salt-affected lands may be used to grow Non-conventional crops of economic value and also such food crops as pearl millet. It is desirable to choose species well suited to saline habitats and to calculate the most economical means of reclamation to make the salt-affected soils productive. The essential ingredients of technology for meeting these problems consist of the use of tolerant species, special planting techniques, and aftercare.

Cultivation of salt-affected areas with palatable halophytes is one of the most promising and ecologically safe approaches in the reclamation process. It also helps cattle breeders and farmers to improve a chronically stagnant economy. Selection of the most suitable halophytic species for introduction into saline land needs extensive research. Malcolm and Sen *et al.* have produced a guide to the selection of salt-tolerant shrubs for forage production from saline lands in southwestern Australia and India, respectively.

Important selection parameters include:

1. Growth and survival for a sufficient period in a representative environment
2. Reproduction by seed or vegetative means
3. Acceptable growth form for management use
4. Production of biomass of sufficient quantity, quality, and acceptability to livestock
5. Ease of establishment
6. Persistence under a profitable management system
7. Effectiveness for erosion control, lowering ground water, and improvement of habitat for wildlife

In addition, the plant must be evaluated by another set of criteria before attempting its development on a crop scale.

1. Establishment:
   a. Seed germination percentage
   b. Vegetative propagation
   c. Seedling vigour and root establishment
   d. Need for supplemental water and nutrition
2. Hardiness under crop production densities:
   a. Insect and disease resistance
   b. Intra- and interspecific competition
3. Ecological traits:
   a. Ecotype variability from which to select stock for introduction
   b. Total genetic plasticity to different ecosystems

Many halophytic species appear to have significant economic potential for desert agriculture. In addition, the productivity of cultivated halophytes is high.

Haloxeric species of the genus *Atriplex* are widely used as fodder crops in otherwise unusable saline wastelands in many parts of the world. Many *Atriplex* species are promising in the reclamation of the salt-affected lands. Use of salt-affected soils for uncontrolled grazing, subsistence cropping, or intensive fuel gathering results in degradation of the natural vegetation cover. This process may take decades to reverse, and the land may never be returned to its original condition.

To slow such deterioration, new economically useful exotic species can be introduced in these areas. Forage-yielding xerohalophytes such as *Atriplex* can be suitable candidates for the management of saline wastelands because these plants can also be irrigated with brackish water. Land reclamation and rehabilitation in arid zones can be achieved by using salt-tolerant plant species for a number of different purposes suited to the local conditions.

Many halophytic species (*e.g.*, *Arthrocnemum* spp., *Nitraria retusa, Salicornia* spp.) are capable of forming adventitious roots on their twigs. This ability varies among species and according to the season of the year. Vegetative propagation is of great advantage in revegetating salt-affected soils. It favours more assured establishment in the field than direct seeding or seedling transplantation. Rooted stem cuttings of *Atriplex* are also helpful in raising a large number of plants with such desired properties as favourable growth habits, regeneration capacity, leafiness, and palatability. Vegetative propagation of desert shrubs is a means of producing genetically identical individuals in species whose sexually produced offspring normally exhibit higher variability. Reduced variability of plant materials can increase experimental precision, and many genetically identical individuals are necessary for varietal testing. Reproduction of desirable parental characteristics such as high seed yield would be valuable in the establishment of seed nurseries. Vegetative propagation is also a method of producing transplants of species whose seeds do not germinate readily.

VEGETATIVE PROPAGATION OF SALTBUSH (ATRIPLEX SPP.)

A. amnicola Paul G. Wilson (river saltbush or swamp saltbush) shows a remarkable high growth rate under desert conditions. The seedlings can be transplanted in the first week of October and can be irrigated with poor quality water. For the first 2–3 months the growth rate is slow, after which fast growth occurs. Enormous production of side branches during the winter season is a very distinctive feature, and these newly formed branches (stems) are soft, fleshy, and purplish pink in colour. By mid-December plants attain a height of about 60–70 cm and lateral branches measure about 50–60 cm. Plants may show two types of growth patterns: (1) an erect type and (2) a prostrate spreading type. Two-year-old plants may cover an area of more than 2–5 m^2. These plants grow sideways and cover the ground very rapidly. Rooted cuttings of *Atriplex* species are needed to establish a rapid plantation. Some *Atriplex* species are subdioecious, with at least three genders. Moreover, rooted cuttings can be used to propagate superior individual plants for a variety of purposes, including breeding programmes and provision of superior or uniform outplanting stock. Observations made in the field have revealed that *A. amnicola* plants have a natural ability to produce rooted cuttings. During the monsoon season, *A. amnicola* was found to produce nodal roots from the lateral branches wherever they touched the ground. This ability is of great importance in binding the loose topsoil. It also helps the plant to recover speedily from grazing pressure and enables the plant to spread rapidly and multiply. Vegetative propagation is much easier in *A. amnicola* because its nodal root formation helps in the production of a large number of rooted cuttings for field planting. The effects of different growth regulators used on stem cuttings for root regulation and axillary shoot growth in different seasons.

Indole Acetic Acid (IAA)

Observations regarding the effect of indole acetic acid on root and shoot growth branch initiation and growth in the rainy season was observed. Interestingly, at almost all concentrations, very large numbers of roots were also produced on the internodal region. Indole acetic acid did not produce much beneficial effect on root and shoot growth; it promoted roots only when administered in lower concentrations. In higher concentrations (40 and 50 ppm) during winter and at all concentrations in rainy seasons, root and shoot growth were affected severely: there was no root formation. IAA favoured root growth only in lower concentrations (10 and 20 ppm) during the winter and summer seasons, respectively. Slight yellowing and drying effects on leaves were seen at higher concentrations.

Indole Butyric Acid (IBA)

The effect of IBA on rooting is next to that of NAA; that is, IBA promotes root growth better at lower concentration (30 ppm) than at higher ones and no distinct difference in the growth of axillary branches was observed. Root growth was maximum in winter at 30 ppm and with welldeveloped secondary roots. Very poor growth of roots and no initiation of axillary branches were observed in plants treated with IBA in summer.

Field Transfer and Establishment of Rooted Cuttings

The effect of growth regulators on root and shoot growth was observed by growing the cuttings in polyethylene bags for 35 days after treatment. It is clear from the results that root growth was maximum at the higher concentration (20 ppm) of NAA, followed by IBA (10 ppm), and the least growth was obtained with IAA (10 ppm) after 35 days. In the control set, the roots were very much shorter than in the treated cuttings. The maximum development of roots with profuse secondary roots was observed with NAA and IBA. Whereas IAA suppressed the growth of roots and axillary branches during summer.NAA and IBA enhanced the growth of axillary branches to a maximum, but the number and the length of the roots were diminished in comparison with NAA. The maximum number of axillary branches was observed in winter and rainy seasons, the least in summer. The propagation of stem cuttings of several saltbush species and a few species from other salt desert shrub genera was studied by Nord and Goodin, Wieland *et al.*, Ellern, and Wiesner and Johnson.

Although Nord and Goodin and Ellern observed a general trend for better rooting of saltbush (*Atriplex*) species in spring than in fall, no data were available for summer and winter. Nord and Goodin noted better rooting of green stem tips than ripe wood cuttings, but Ellern failed to find any difference in rooting of soft, green cuttings and young woody stem cuttings.

Nanda *et al.* used IAA, IBA, and NAA to enhance the rooting response of stem cuttings of forest trees and investigated the possibility that even seasonal changes in the effectiveness of different auxins are governed by morphophysiological factors. Auxins enhanced the rooting of stem cuttings of *Populus nigra* and *Hibiscus rosa-sinensis* even during December–February, but these hormones failed to cause rooting in *Ficus infectoria* cuttings during the same period. It was observed that auxins enhanced the rooting more in winter, followed by the rainy season, and least in summer.

Indole acetic acid has been one of the most commonly used auxins, but different workers have obtained varying results. Chatterjee found that *Pogostemon potehouli*, an essential oil–yielding plant, responded more favourably to IAA than other auxins. Shanmugavelu also obtained the maximum percentage of rooting in cuttings of certain shrubby plants with IAA. On the other hand, NAA gave favourable results in the induction of roots in cuttings of *Levendula, Ficus infectoria*, and *Hibiscus rosa-sinensis*. The experimental results of our study showed that a large number of roots were produced at lower concentrations of NAA, IAA, and IBA.

A number of saltbush species may be established from cuttings, including *A, amnicola, A. nummularia, A. canescens, A. halimus, A. lentiformis, A. paludosa*, and *A. polycarpa*. The cuttings should be taken at the peak of spring growth or in the autumn in a Mediterranean climate. The wood should be about 6 mm thick and 250 mm long, taken from young stems between two leaf axils. A rooting hormone (*e.g.*, IBA) may be applied to encourage root growth before approximately half the stem is covered with a moist, sandy soil. The cuttings should root within 6 weeks and should be ready for transplanting in 10 weeks. In our study, IBA also enhanced the rooting in *A. amnicola*. According to Richardson *et al.*, fourwing saltbush cuttings could be rooted best in the summer, but *A. amnicola* rooted best in winter, followed by the rainy season and summer.

According to Sharma and Sen and Rajput and Sen, respectively, winter is most suitable for the vegetative propagation of *Tamarix* and *Atriplex*. The present results also support these views.

The results of field experiments showed that NAA is more effective than IBA and IAA. The increased appearance of new leaves with an increase in the percentage of rooting also points to better rooting possibilities, with the emergence of more new leaves on the cuttings. The greater number of roots per cutting and the greater number of leaves may also help the cuttings to survive when sown in natural conditions.

VEGETATIVE PROPAGATION OF MULBERRY (*MORUS* SPP.)

Since the dawn of agriculture, one of the principal aims of human beings has been the control and promotion of plant growth to satisfy human needs. These two important aspects of people's work with plants in the struggle to increase production are by no means synonymous. Humans soon realized that lush green growth does not always produce the best crop in the form of fruit and seeds, and hence they were forced to evolve such well-known cultural methods as pruning, balanced manuring, and use of mineral fertilizers to regulate the nature and luxuriance of plant growth.

The naturally occurring (endogenous) growth substances are commonly known as plant hormones, while the synthetic ones are called growth regulators. A plant hormone (synonym: phytochrome) is an organic compound synthesized in one part of a plant and translocated to another part, where at very low concentrations it causes a physiological response. Plant hormones are identified as promoters (auxin, gibberellin, and cytokinin), inhibitors (abscisic acid, xanthoxin, and violaxanthin), and ethylene and other hypothetical growth substances (florigen, death hormone, etc.). They usually exist in plants and crops at a concentration lower than 1 M; above this, they are generally considered supraoptimal.

Mulberry is propagated either through seeds or vegetatively. The latter is the more common method of propagation because of such advantages as maintenance of particular properties of the plant, relative speed in raising saplings in large numbers for plantation, adaptability to a particular habitat, and abilities to develop resistance to pests and diseases and to modify the growth of plants. Propagation through seeds has reached certain limitations. For example, triploid plants, which do not produce viable seeds, cannot be propagated. It is not possible to reproduce true to the type from a seed of biparental origin. Mulberry is a highly heterozygous plant that is open for cross-fertilization. Therefore, the seeds that are formed through open pollination are natural hybrids.

Seedling populations from such seeds provide wider chances for selection of superior types whose characteristics are perpetuated through vegetative propagation. Generally, the population thus obtained is a mixture of several clones. Each clone is heterozygous although homogeneous, and the same genotype is maintained because propagation is vegetative, Interclonal variations are due to heredity. Depending on climatic and soil conditions, different countries follow different modes of vegetative propagation. Hamada described the methods used in Japan, which include (1) bark grafting (Fukurotsugi), (2) veneer grafting (Kiritsugi), (3) simple layers (Magedori), (4) continuous layers (Shumokudori), and (5) division (Shirodasmi), hardwood cuttings (Kojyosashiki), and softwood cuttings (Shinshosashiki), Generally, grafting is used in places where the temperature is 6°C in March and more than 25°C in July, with rainfall of 175 mm. Shirodasmi cottage is popular in places having temperatures less than 4°C in March and less than 25°C in July with rainfall lower than 175 mm. Propagation through hardwood and softwood cuttings is common in the northern districts and the southern region, respectively, of Japan. In Italy, rooted grafting is a popular method of multiplying Japanese mulberry varieties.

In India, the most common method of propagating mulberry is through cuttings in multivoltine regions (*e.g.*, Karnataka and West Bengal). Exotic varieties that are not established by cuttings are propagated through root grafts. Many of the indigenous varieties and well-acclimatized exotic varieties are propagated through cuttings. Bud grafting (budding) is used only when scion material is scarce. Whenever a large number of mulberry plants must be obtained in a shorter time than would be possible if they were started as a cutting, the method of layering is used. Layering allows the grower to fill in the gaps formed as a result of the failure to sprout of certain cuttings planted in pits of established plantations.

In univoltine areas (*e.g.*, Kashmir), the mulberry is propagated through seedlings and the exotic varieties through root grafts. In India, the field-scale propagation through cuttings of Japanese varieties of mulberry is still a problem. Propagation through seeds is used mainly to bring about a varied population for the purpose of selection and hybridization. Because mulberry flowers are open for cross-pollination, the seeds thus collected serve mainly as sources of stock material for grafting. In general, a deficiency of hormone must be created experimentally (as by removing young leaves or using a hormone-deficient mutant) to show that adding a hormone has an effect.

In this respect, the Mitscherlich law of diminishing return can be modified as follows: the increase in plant response produced by a unit increment of a deficient (limiting) hormone is proportional to the decrement of that hormone from the maximum. Mulberry varieties that do not ordinarily produce roots from a cutting are induced to root with application of the requisite quantity of root hormones. The following chemicals are generally used, but their efficiency varies from species to species and from variety to variety: (1) IAA, (2) IBA, and (3) NAA. The objective of using growth regulators is to increase the percentage of cuttings that form roots, hasten root initiation, and increase the number of roots per cutting. IBA and NAA have proved to be better in producing roots than other growth regulators. The water requirement of mulberry does not differ greatly from species to species or from variety to variety. The plant must be capable of absorbing water from soils of low moisture regimes. Generally resistant plants should have well-developed root systems, hydrophilic colloids to absorb and hold water by imbibition, and adaptations to facilitate the lowering of transpiration. In this regard, certain Japanese varieties have a thick cuticle, sometimes a two-layered epidermis, a palisade parenchyma, and other beneficial characteristics.

Although many tropical species root profusely through cuttings, certain temperate varieties do not ordinarily produce roots. Root induction has been successfully achieved in the latter varieties by the (artificial) application of the requisite quantity of root hormones. However, the efficacy of the substances varies from species to species and from variety to variety. Development of the root primordium depends on the relative amount of natural auxin present in the plant. Varieties that do not root apparently contain less auxin. The growth regulators act like auxins when applied in small quantities and move upward in mass translocation through the xylem when the bases of the cuttings are soaked in their solutions. The objective of treatment is to increase the percentage of cuttings that form roots, hasten root initiation, and increase the number of roots per cutting. Indole butyric acid and naphthalene acetic acid appear to be better at producing roots than other agents. The chemicals may be applied by various methods, including direct application of a powder, soaking the cuttings in dilute solutions, dipping the cuttings in concentrated solutions, and application as a paste in lanolin.

The action of many gibberellic acids (GAs) is similar to that of IAA, including elongation, promotion of cambial activity, induction of parthenocarpy, and stimulation of nucleic acid and protein synthesis. The GA3s vary greatly in their biological activity, and GA5 and GA7 are considered to have the widest range. In ferns, algae, and fungi, GA3s have also been shown to influence growth and development. For the vegetative propagation experiments, mulberry cuttings were collected from both cultivated and wild varieties at Jodhpur (site Chopasni). Growth regulators used for root initiation in cuttings were NAA, IAA, IBA, and GA. Shoots of thick branches with well-developed buds were used for rooting experiments. Cuttings taken from parts with a high carbohydrate content have been reported to root more readily and profusely than cuttings selected from parts rich in nitrogen. Portions of the shoot that were too tender at the top and overmature at the base were rejected. Cuttings taken from young branches sprouted rapidly and profusely as compared with those taken from old parts. Cuttings of 7 to 10 cm usually of pencil thickness with three to four well-developed buds were prepared from the central portion of the clone with a slanting cut. Also, the total leaves generally increased, together with the number of inflorescences. Increasing concentrations of hormones tended to decrease the values. Slightly higher values of these parameters were observed with 10 ppm than with 20 ppm IAA. Of the two auxins, IAA was more effective than NAA. The case of NAA, a lower concentration is more effective than a higher one. From the observations of the rooting behaviour in a wild variety of mulberry, we see that the lower concentration of IAA is more effective than the higher one. The maximum number of sprouting buds was 11; afterwards the values remained constant. However, in the case of NAA, 20 ppm was more effective than 10 ppm. Comparatively, IAA was more effective than NAA and higher values were observed in the wild than in the cultivated variety.

Cuttings were immersed in different concentrations of growth regulators for 24 hr. During treatment, cuttings were kept inside the growth room. After this treatment, cuttings were washed in distilled water and individually transferred to test tubes filled with water. Three cuttings were used for each set, and each set was repeated three times for confirmation. The observations were recorded after a definite interval of time. Measurements of bud sprouting, number of leaves, inflorescence, and root initiation were observed in the cultivated variety. The lower concentration (10 ppm) of IAA led to a better response than the higher one (20 ppm). At 10 ppm the bud sprouting was 100 per cent, whereas with 20 ppm it diminished to 4. Leaf initiation and inflorescence followed the same trend as the bud sprouting, being 8 and 11, respectively, in 10 ppm, and 8 and 6, respectively, in 20 ppm.

Similar results were obtained in NAA, with 100 per cent bud sprouting in 10 ppm, slightly more than in IAA, being 5. Leaf initiation and inflorescence were also higher in 10 ppm compared with 20 ppm, being 9, 9, 8, and 5, respectively. IAA again showed a beneficial effect at the lower concentration (10 ppm) as compared with the higher (20 ppm), producing 7 and 4 buds, respectively. Leaf initiation increased from 7 to 10 with increasing concentration, but the inflorescence did not show any change. The results with NAA showed effects similar to those with IAA. At 10 ppm, 8 buds sprouted out of 8 buds, whereas at 20 ppm the figures were 4 out of 7 buds. Leaf initiation showed a better response at the lower concentration than at the higher one, and a similar trend was also shown for inflorescence. Growth means an irreversible increase in the weight, area, or length of a plant or a particular tissue or organ of a plant, while development denotes the changing pattern of organization as growth progresses. Control over plant growth by the regulated exogenous supply of chemical substances may occur in different ways. It has become clear that total control of plants is vested not in a single hormonal type; rather, control is shared by a group of several specifically defined auxins, gibberellins, ethylene, and certain naturally occurring inhibitors such as phenols and abscisic acid. Thus, the plant growth regulators provide a very helpful tool for controlling physiological processes in plants. NAA was found to be better than IAA in rooting by Jauhari and Rehman in cuttings of sweet lime. It responded favourably on induction of roots in stem cuttings of many plants. In the present study IAA was found to be more effective than NAA.Stem cuttings of *Ipomoea pes-caprae* and species of *Morus* showed a large number of roots and buds in the higher concentration but with maximum suppression of growth, whereas lower concentrations resulted in only improvement in the growth of roots. In our investigation also, the higher values were observed with lower concentrations of the growth regulators. Under favourable environmental conditions, during the period of root development, a callus tissue develops at the basal end of a cutting: an irregular mass of parenchyma cells in various stages of lignification. Callus growth arises from cells and adjacent phloem, although various cortical and medullar cells also contribute.

Because root development and callus formation occur simultaneously, it is believed that the formation of callus is essential for root development. In reality, these two are entirely different phenomena. Sometimes roots develop even without callus from the nodes. Callus formation is sometimes beneficial in varieties that are slow to root because it provides a protective layer, preventing the cutting from becoming desiccated and decayed. Sometimes the callus interferes with the absorption of water by the cutting. In our investigations, rooting did not start, instead, callus formation was observed after 1 week of treatment. The callus was creamy white and had a granular texture.

The rate of sprouting of vegetative buds is of primary consideration in introducing a variety or species in an area. Mulberry varieties grown in Mysore and West Bengal sprout throughout the year, facilitating the attempts of sericulturists to rear the silkworms year-round. The axillary buds vary in size, shape, and position from variety to variety. Thus the rooted stem cuttings are more useful than seed sowing because the survival of a rooted cutting is far better than that of the delicate seedlings in the stressful environment of the desert.

SEED STORAGE IN CERTAIN FRUIT CROPS

Mango

Mango seeds are protected by hard stony endocarp. Seed is ex-albuminous having two fleshy cotyledons with testa and tegmen represented by two papery layers. Polyembryony is observed in certain cultivars. Seed exhibits recalcitrant type of storage behaviour wherein seed loses the viability on drying after extraction and undried seeds cannot be preserved long at low temperatures. Reduction of seed moisture to below 25 per cent in Alphonso and 32 per cent in Totapuri lowered the germination percentage and seedling vigour (Doijode 1990). Patil *et al.* (1986) obtained 40 per cent of germination in Alphonso stones after 90 days of storage at 25°C in polyethylene bags with charcoal powder. It declined from 75 to 12 per cent after 60 days when stored in open while those stored in polyethylene bags at 8°C did not germinate at all. The moist storage appears to be successful for preservation of recalcitrant seeds. Here seeds are imbibed with either germination inhibitor or fungicides for preventing germination and pathogens during moist storage. Seed treatment with 1 per cent 8-hydroxy quinoline sulphate and storage at 15°C could preserve the seed viability for one year (Doijode 1995).

Citrus

Citrus seeds are either mono- or polyembryonic in nature. They are largely Utilized for raising rootstocks as well as in breeding programme. Lime (Citrus aurantifolia) seeds are small, oval and polyembryonic with white cotyledons. Lemon (Citrus limon) seeds are ovoid and polyembryonic with 10-15 per cent nucellar embryos and white cotyledons. Sweet orange (Citrus sinensis) seeds are obovoid, polyembryonic with white embryos. Mandarin (Citrus reticulata) seeds are small polyembryonic with green embryos. Pummelo (Citrus grandis) seeds are large ridged yellowish and monoembryonic. Grape fruit (Citrus parodisi) seeds are white, polyembryonic with white cotyledons. Seeds of many Citrus species are viable for a short period under ambient conditions.

The seeds belong to orthodox group and/or show intermediate storage behaviour and require special procedure for preservation. Seed storage in moisture proof containers at low temperature checks the deterioration by reducing metabolism. Seed storage with calcium chloride in polyethylene bags at 8°C resulted in 100 per cent viability in Citrus karna and rusk citrange and 84 per cent in Citrus jambhiri and Citrus limonia after 150 days (Krishna and Shankar 1974). Acid lime seeds were successfully stored for 10 years at 5 and -20°C. Grape fruit seeds retained viability for 3 months when stored with charcoal at room temperature and for 4 months at 5-8°C (Chacko and Singh 1968). In Poncirus trifoliata and Citrus grandis, germination was the highest after 80 days in seeds stored in polyethylene bags at 4.4°C and 56-58 per cent RH (Mallareddy and Sharma 1983). Seeds could be stored in fruit itself for short term preservation. Fruits of some lemon cultivars were successfully preserved for 40 days at ambient temperature and for 60 days at 5°C (Doijode 1984). Bajpai *et al.* (1963) also reported that seed storage in whole fruit of kagzi lime and sweet orange is better than in polyethylene bags.

Rambutan

Seeds are difficult to store under normal conditions (Ellis 1984) and at room temperature they lose their viability within a matter of days. They can be stored for a few weeks in moist saw dust or charcoal to which some aril juice has been added at 21-28°C.

Jackfruit

Seeds are large oblong about 3 x 2 cm in size having thick gelatinous yellow covering and belong to recalcitrant group. Seeds remained viable for 15 days and gave 40 per cent germination after 30 days of storage.

Litchi

Litchi seeds are dark brown, covered with white fleshy, juicy and translucent aril. Seeds are recalcitrant and should not be allowed to dry. When sown immediately after extraction, their germination was 100 per cent which reduced just after one day storage in water and it was 75 per cent on the seventh day and 30 per cent on 15th day after storage. Seeds stored in moisture proof containers sprouted within 15 days and those stored for 10 days in polyethylene bags showed 50.7 per cent germination. However, seeds can be preserved in fruit for short periods. Seed viability was preserved for 24 days by storing fruits in sealed polyethylene bags after treatment with benomyl (0.05 per cent) and 6 per cent wax emulsion.

Mangosteen

Seeds are recalcitrant and lose their viability rather quickly, especially when the thin membrane around the seed is damaged or when seed is placed in ambient or cool atmosphere. Storage of seeds in charcoal or moss at room temperature retains the viability for a few months.

Conclusions

Seeds of several tropical fruit crops show recalcitrant storage behaviour and they do not withstand drying or are able to keep alive at low temperatures. Thus, they are difficult to store for longer period. In this context, more research is needed for extending storage life. However, it is desirable to preserve the germplasm in field as fruit repository. To supplement the conservation process, wherever, as far as possible seeds could be conserved by use of available techniques like moist storage, etc., for certain period.

GERMPLASM OF MANGO

Mango is one of the most important fruit crops in India. It has been in cultivation for almost last five thousand years. Apart from India, mango is also an important crop in the Philippines, Indonesia, Thailand, Myanmar, Malaysia, Sri Lanka, U.A.E., South-East Africa, South Africa, U.S.A., Mexico, West Indies, Brazil and tropical Australia. All the cultivated Indian mangoes belong to the single species Mangifera indica L. The geographical distribution, phytogenetic trend, pollen morphology, chromosome number, cytogenetical and breeding behaviour indicate that the highest concentration of species of Mangifera is found in Malayan Peninsula followed by Sunda Islands and the Eastern Peninsula comprising Myanmar, Thailand and Indo-China. Occurrence of wild species like M. sylvatica and M. caloneura in Myanmar, Assam and northeastern India and evidence of fossil leaf impression of M. pentandra which looks similar to M. indica having edible fruits indicate that mango originated in this region. In India, more than thousand varieties of mango are available and these have arisen mainly as a result of open pollination.

It is to be admitted that most of the varieties lack in one or the other character. As opined by Naik *et al.* (1958), this very diversity has tended to limit commercial production of fruits of standard quality in India. In the recent past, efforts were made to exploit natural variability and breed new varieties for different purposes. So there has been a need to prevent the loss of varieties, and to conserve and use them in the breeding programme. Apart from intervarietal breeding programmes, progress has been made with regard to identifying the location of wild species of Mangifera and their utilization in breeding.

Collection and Conservation

Mango in India has a long history of domestication and cultivation. The rich varietal wealth, however, has not been collected at one place. Hence, there is erosion of the genepool owing to deforestation, urbanisation and industrialization. Systematic attempts to conserve the genepool have been made in the past few decades and more than 1266 varieties were collected in various field genebanks.

In-situ Conservation

The species of Mangifera occur mainly as a complex biotic community in tropical humid forests, subtropical rain forests and dry forests/woodlands of Indo-Malayan biogeographic realm. For in situ conservation, the hilly regions of east Orissa, and forests bordering Myanmar in Manipur valley have been identified.

Ex-situ Conservation

Most of the important commercial varieties grown in India as stated earlier have arisen due to natural cross pollination and selection. Germplasm collections have been made in the past. Although, substantial diversity of cultivated M. indica is in the field genebank, this does not contain wild species. Germplasm collections at many centres are duplicated and it has been noticed that varieties belonging to one region behave differently in other regions. Yadav and Singh (1985) opined that south and north Indian varieties belong to two different ecotypes of M. indica. Mustaffa (1993) confirmed the same by isozyme analysis. Hence, diverse collection has become a necessity to get actual performance. The major germplasm collection centres are listed.

Conservation in Vitro

One of the main constraints in the maintenance of mango in field genebank has been that it requires large area and funds. In this regard, in vitro conservation holds promise. However, in mango, owing to the problem of regeneration of tissue, so far no success has been obtained. Iyer and Subramanyam (1989) have suggested pollen storage of important mango varieties. This will also help in speeding up the hybridization programme.

Utilization

Almost all the commercial cultivars belong to M. indica. However, a few commercial cultivars of South-East Asia belong to other species, such as M. altissima, M. caesia, M. foetida, M. griffithii, M. odorata, M. pentandra, M. sylvatica, M. zeylanica, M. laurina, M. lagenifera, M. cochinchinensis, etc. For the improvement of mango, some of the wild species could be of great help, Iyer (1991) suggested that Mangifera species useful for crop improvement fall into two categories: (i) species having edible fruits with other desirable characters, and (ii) species that could act as gene donors for specific improvement like resistant to pests and diseases. If basic information like crossability become available then interspecific hybridization in mango can become a reality.

CONSTRAINTS IN DECIDUOUS FRUIT PRODUCTION

Large number of old orchards (more than 30 years old) are showing decline in terms of growth and fruit yield. Such old trees do not produce adequate extension growth. Large scale replanting is therefore needed. Delicious group of cultivars constitute the major share (about 83 per cent in H.P.) of apple production in the country. These cultivars are self unfruitful and need cross pollination to ensure good fruit set. Interplanting pollinizer cultivars (Golden Delicious, Jonathan, Red Gold, Lord Lambourne etc.) in the proportion of 25 to 33 percent is necessary for good fruit set, and choice of wrong pollinizers and their inadequacy in number often result to low productivity.

In many countries, Delicious group has been replaced or is in the process of replacement with more promising cultivars. The need for injecting new blood into the apple industry through spread of new cultivars (spur types, colour mutants, strains of Gala, Red Fuji; scab resistant cultivars, bud sport selections of Royal Delicious, and some of the promising hybrids) is urgently felt. Some of the spur type and coloured mutants are already popular with farmers and high density planting has also caught the imagination of developmental departments and agencies both in H.P. and JandK. The research system has already identified Early, Mid and Late cultivars for different agro-climatic regions. The low chilling cultivars of stone fruits have also covered large tracts of the sub-tropical plains of Punjab, U.P and H.P. For the hills, promising cultivars identified need further spread. Generally, apple is grown in marginal land and fertilizers are not applied according to the requirements of the trees. The water and fertilizer use efficiency is generally poor. Also, spring frost and hailstorms are adverse weather parameters leading to low productivity. Research results have shown that through proper orchard management practices (soil and water conservation and fertilizer application) the fruit yield can be doubled in the existing orchards.

The adoption of improved production technology developed by the research system can bring visible and perceptible changes in the temperate fruit industry in India. Technologies like use of clonal rootstocks, introduction of renewal pruning techniques and micro nutrient applications have not been transferred and adopted at a satisfactory level. Apple scab disease has been the major plant protection problem in apple in JandK and H.P, whilst U.P hills are comparatively free from the disease. Apple scab forecasting system developed and the chemical control schedule prescribed have helped in reducing loss due to apple scab to a considerable extent. Apple growers are adopting the prescribed schedule of chemical sprays to control the disease. For checking entries of diseased material in the free areas of U.P. and North-Eastern Hills, strict quarantine and selection of elite disease-free mother plants are very essential. Often it is not followed strictly. Some of the virus diseases have also been reported in apple for which biological and serological indexing/detection procedures have been developed. Limited quantity of virus-free budwood is also being supplied. Extreme care is now required to be taken to multiply quality planting material (in apple alone approximately 2 million plants/year) for establishing new plantations.

Most of the orchardists still sell their crop at flowering to contractors as there is no well organized marketing system. Transportation in the hills itself is problematic. Post-harvest management problems originating from poor harvesting (strip picking) and improper packing system (non CFB boxes) and lack of proper pre-cooling and cold storage facilities result in huge (25-30 per cent) loss of fruits. Capacity of the processing sector is also inadequate. Product diversification, value addition and market infrastructure development would require very substantial investment. The existing processing units are quite old and they require modernization for which substantial investment is required. CA storage trials have shown good promise. Its extension in larger growing areas is needed. Technology for storage of apple is now known, as a result of which apple is now available throughout the year.

GOVERNMENT POLICIES AND PLANS

Research on pome and stone fruits is conducted mainly by three State Agricultural Universities, namely: (a) Sher-e-Kashmir University for Agriculture Science and Technology, JandK; (b) Y.S. Parmar University for Horticulture and Forestry, H.P.; and c) G.B. Pant University for Agriculture and Technology,

U.P. A good number of research stations of these Universities located in major pome and stone fruit growing belts are engaged in temperate fruit research.

The Indian Council of Agricultural Research (ICAR) through All India Coordinated Research Projects on Fruits and Post-harvest Technology and another Network project on Apple Scab disease has further strengthened the research activities in deciduous fruits. A few long established temperate fruit research stations namely at Shalimar Bagh in JandK, at Mashoobra in H.P. and at Chaubatia in U.P. hills have made commendable progress in temperate fruits. Both State Governments and the ICAR provide financial support to strengthen fruit research in these Universities. A large number of ad-hoc research projects on pome and stone fruits funded by ICAR are also generating good information on these crops.

During the 8th Government Developmental Plan (1992-1997) the ICAR established a Central Institute for Temperate Horticulture (CITH) with its headquarters at Srinagar in the JandK State, with a regional station at Mukteshwar in the U.P. Hills. Both these research stations will work exclusively on temperate fruit crops. This new Institute will receive major support during the 9th Plan period (1997-2001). In the North-Eastern Hills region, the ICAR Research Complex for NEH Region with its headquarters in Meghalaya State and regional units in each of the 5 other States are engaged in fruit research.

On the developmental side the State governments are engaged in nursery production of quality planting material. For example, in H.P. alone currently there are 600 nurseries in private and public sector producing and distributing more than 0.8 million plants of apple alone, every year. There are ambitious programmes in all the States to further expand/replant with new improved cultivars. Apple scab disease control and post-harvest processing sectors are getting focused attention in Government developmental plans. The Directorate of Marketing and Inspection of the Government of India has framed grade standards for apple, plum and William pear. The organizations like National Horticultural Board (NHB), National Cooperative Development Corporation (NCDC), Agricultural and Processed Food Products Export Development Authority (APEDA) etc., are providing incentives to traders and exporters to improve their infrastructural facilities like grading and packaging centres, refrigerated transport, setting up of pre-cooling, cold storage, auction platforms etc. The NCDC is undertaking procurement and marketing of apple on a limited scale. The NHB has set up a market information service for the benefit of growers. Deciduous fruits, covering pome and stone fruits contribute significantly to the horticulture economy of India. Apple production dominates the scene and systematic cultivation and marketing of apple can change the rural economy in the hills of North-Western India. New vision and concerted efforts are required for change in variety mix, supply of quality planting material from elite clones on indexed clonal rootstocks. High density planting, water management including micro-irrigation, integrated plant nutrient management and IPM strategy for plant protection are some of the areas which need greater RandD focus. Adoption of post-harvest management practices and infrastructure development for grading, packaging, pre-cooling and storage of the produce needs focused developmental attention. Value addition and export promotion, particularly of apple are drawing due attention of the developmental agencies in India.

FRUIT CULTURE

Introduction: The Cashewnut (Anacardium occidentale L.) is one of the important nut tree grown in the tropical world. This fruit tree was introducted into Malbar coast of India in 16th century by Portuguese. It is the native of Brazil. It is grown mostly on West Coast area of Maharashtra, Goa, Karnataka, Kerala and Tamil Nadu.

Importance: The dry nuts - Kernels are of high nutritive value and also rich in protein, carbohydrates, minerals like calcium, phosphorus and iron. This has a great export value and is considered as one of the important horticultural crop in India.

Climate: The Cashew requires a minimum rainfall of 60 cm per annum, but can stand extremes of rainfall from 20 cm to 400 cm. If there is sufficient water supply, it can withstand a long period of dry spell and low humidity. The Cashew is a sun loving tree and does not tolerate excessive shade, also it does not favour very high temperatures above 45 deg cent. During the fruit set and development. Heavy rains and cloudy weather during flowering adversely affect the yield.

Soil: The Cashew is cultivated on a wide variety of soils. It is considered a crop of marginal land and is recommended for slopy and light soils. The best soils for better production are deep, friable, well drained and without a hard pan upto 2/3 m in depth. The Cashew is mainly grown on Laterite, red and coastal sands in India.

Varieties: The following are the important varieties of Cashew grown in different parts of the country.

i. Vengurla 1
ii. Vengurla 2
iii. Vengurla 3
iv. Vengurla 4
v. Vengurla 5
vi. Vengurla 6
vii. Bhubaneshwar
viii. Kanaka
ix. Dhana
x. Selection

Propagation: The cashew is grown by the following methods:

i. *Seed Propagation:* It is the oldest and cheapest method of propagation. It is also used to raise the plants for the purpose of grafting.

ii. Layering this method is more successful in coastal and more humid areas.

iii. *Soft Wood Grafting:* This is followed both in the nursery as well as in the field of in-situ. For this purpose, veener grafting is done.

iv. *Top-working:* This method is used not for propagation but for converting inferior free into desirable one.

Planting and Season: Normally cashew grafts are planted at the spacing of 7.0 x 7.0 or 7.5 x 7.5 or 8.0 x 8.0 m on plain land or on desirable contours on slopy lands. After marking, the pits of 60x60x60 cm are dug and filled with a compost,top soil, single super phosphate and karnaj cake. This is to be well before the monsoon starts. Planting should be done during June-July *i.e.* in the beginning of monsoon.

Interculturing:

i. Removal of weeds is done once twice

ii. Intercrops are planted in interspace

iii. Some perenial forest plants and/or some seasonal crops can be taken as per the local need and situation.

Care of young orchard:

i. Gap filling for the missing plants

ii. Removal of outgrowth on stocks in case of grafted plants

iii. Staking with bamboos.

iv. Cover cropping

v. Providing protective irrigation during first few summers. These are some of the points for young plantations.

Special Horticultural practices:

i. Pruning of dead and dried shoots alongwith crisscross branches and water shoots

ii. Pruning of leader shoots in June followed by 2 per cent KNO_3 spray

iii. Spraying with 10 ppm NAA twice during flowering for increasing fruit set and minimizing flower and fruit drop.

Irrigation: The cashew is mostly grown as the rainfed crop and requires no irrigation in the high rainfall areas. However, if the rainfall is low, the cashew responses well to irrigation water at the time of fruit set upto full development stage of nuts. Irrigation should not be given before or at the time of flowering, as it is likely to promote vegetative growth.

Nutrition: To improve the growth and yield the cashew trees should be manured and fertilized with bio-fertilizers like biomeal and NpK 500,200, 200 gm per tree per year. Spraying with ultrazyme and 8:12:24:4 NPKmg twice, during fruit development is useful to increase size and yield of nuts.

Plant Protection: The crop particularly blossom and fruits should be protected against Tea Mosquito, Stem and root borer is also an important pest. Leaf minor, Leaf and blossom webber, Flower thrips are the common pests in neglected orchards. Inflorescence blight, pink disease, and anthrac nose are some diseases found on cashew. Clean cultivation, pruning, better aeration and preventive measures should be adopted besides occassional sprays of selected pesticides and fungicides.

Harvesting and Yield: The Cashew comes to its full bearing stage at the age of 10th year and continuous to give yield further for 30-40 years. Usually the nuts are collected after they fall off from the tree. They are separated from apples and dried for 2/3 days to reduce the moisture. Regarding yield there is a wide variation and differences of opinions. However, the desirable variety at good management yields 50 kg of raw nuts.

Post Harvest Handling: The processing of raw nuts involves roasting, shelling, drying, peeling, grading and packing `A' grade nuts are always in demand from the outside markets and fetch a good price. A Fenny like drinks are prepared from the apples.

FOOD PRODUCED IN GREENHOUSES

Although greenhouses have been in existence since 1800 (or earlier), and greenhouse food production started to develop as an industry in the second half of the nineteenth century, the largest growth and expansion of the greenhouse industry occurred throughout the world following World War II. Today, food production in greenhouses can be found in all continents. Most popular food crops grown in greenhouses are tomato (beefsteak, cluster, Italian, cherry), cucumber, and sweet pepper.

Other greenhouse grown vegetables include watermelon, muskmelon, summer squash, zucchini, lettuce, eggplant, snap beans, celery, cabbage, radish, Welsh onion, and asparagus. Fruits such as grapes, strawberry, banana, pineapple, papaya, orange, mandarin, cherry, and fig, as well as culinary and medicinal herbs, are also grown in greenhouses.

TODAY'S GREENHOUSES

Covering materials: The main greenhouse covering materials are glass and polyethylene (PE). Glass has been used since the early days of greenhouses. The introduction of PE film after World War II was the main reason for the expansion of greenhouse production around the world, and it is now the most widely used covering material in the world. Glass-covered greenhouses are concentrated mainly in northern Europe and North America. The low cost of the PE greenhouse is the main reason for its high popularity, especially in developing countries. In recent years, the use of PE-greenhouses has even spread to northern regions. Research has shown that, under Canadian climatic conditions, heating costs of a double-layer PE-greenhouse are 20 to 30 percent lower than for a glass-covered greenhouse. Most of the greenhouses built now in Canada are covered with PE. Standard PE film blocks the ultraviolet, but not the infrared radiation, and has a short durability. However, improved PE films retain the infrared, but allow the ultraviolet, radiation (necessary for the bees, used for pollination of plants, to orient themselves) in the greenhouse, and are more durable. Polyvinyl chloride (PVC), another plastic film used to cover greenhouses, is used mostly in Japan. Other covering materials for greenhouses include rigid plastic acrylic, fibreglass, polycarbonate, and PVC panels, but their use is generally limited because of their high cost, compared to PE. Beside glass and PE, polycarbonate is often used on the sidewalls of polyethylene greenhouses in northern regions because of its good insulation, durability, and reasonable cost.

Technology in the greenhouse: Greenhouses come in many styles and sizes, from the original houses with minimal climate control (furnace and vents) to the modern 10-ha (25-acre) or more, multispan greenhouses with high-tech climate controls (sophisticated and powerful heating system, CO_2 enrichment, evaporative cooling pads, exhaust fans, roof vents, thermal/shade curtain, computer controls, light sensors). Most sophisticated greenhouses are generally found in the developed, northern countries. Phytotrons are highly sophisticated structures that allow for accurate control of environmental conditions including light, and are generally used for scientific research in universities and research institutes. However, phytotrons cannot be considered greenhouses since they are not covered with a transparent material.

The degree of environment control needed depends on various factors. The first factor is the location of the greenhouse (local climatic conditions). Northern regions are characterized by cold winters and warm summers. If the objective is to grow plants all year long, then such large differences in climatic conditions between winter and summer require a high-tech greenhouse. In regions such as the Mediterranean (Spain, Italy, Morocco, Greece), the mild winter climate does not require the use of powerful heating systems, and low-tech greenhouses are sufficient for winter production. However, these regions have very hot summers, and the use of a low-tech greenhouse may not provide satisfactory temperature control to grow plants during summertime.

The production schedule also affects the level of environment control and thus the level of technology. A greenhouse in northern regions may require a high level of climate control if the objective is to grow crops all year long (or long-season crops). If the objective is only to extend the production season (*e.g.*, one early crop in spring), then a less sophisticated greenhouse could be satisfactory. Optimal growing conditions differ from one species to another. For example, lettuce prefers cooler temperatures than cucumber. Thus, the crop grown in the greenhouse may influence the level of environment control needed or desired. A low-tech greenhouse may provide sufficient climate control for lettuce but not for cucumber, depending on the location of the greenhouse and the production schedule. Economic development also plays a role in the level of technology used in the greenhouse. In developing countries, growers may not be able to afford the most sophisticated equipment, and may lack technical expertise and technical support.

Greenhouses in desert regions: Although greenhouses were developed in northern regions as a means of protecting crops against cold temperatures, and are therefore generally associated with cold climates, they are also used in arid regions such as Saudi Arabia. In such regions, the objective of the greenhouse is to protect plants from the excessive solar radiation and temperature, and to prevent excessive water loss by plants (especially since water resources are generally limited in those regions). Therefore, technology in greenhouses in these regions is directed towards cooling.

Artificial lighting: In northern countries, high-tech greenhouses can provide optimal growing conditions (temperature, humidity, carbon dioxide) for vegetable crops even during the coldest winter months. However, even with excellent climate control, yield and quality of crops grown during these months are low due to the low light level available. Research has shown that it is possible to produce good yield of high-quality produce during the winter months by using artificial light to supplement the natural radiation. The most common artificial lighting is the high-pressure sodium lamp. The high cost of electric energy in many regions is the most important factor preventing an increased use of artificial light.

PRODUCTION SYSTEMS

Growing in soil: Since the early days of greenhouses, plants have been grown in soil or in soil-filled containers. The first technique for fertilizing plants, which is still in use today in organic production, was the use of manure. Today, fertilization of plants can also be accomplished by incorporating chemical fertilizers in the soil, or by distributing fertilizers dissolved in water (so-called fertigation) to plants with a drip (trickle) irrigation system. Intensive and repetitive cultivation of crops on the same soil generally results in a degradation of soil properties and fertility.

Salt accumulation may be another problem in soil cultivation. Incorporation of manure, compost, and other organic materials into soil can be used to improve its structure and replenish its fertility. However, ensuring perfect fertilization of plants grown in soil is still a difficult task. Furthermore, intensive and repetitive cultivation of crops on the same soil can also result in insect or disease infestation. Soil replacement and soil fumigation are two solutions, but the first technique is expensive and the second is not always successful. Greenhouse production in soil is still used widely.

GROWING WITHOUT SOIL

In order to better control fertilization for optimizing plant growth and yield, and also to avoid the problems occurring in soil, growing systems that do not use soil (soilless) were developed for the cultivation of greenhouse crops. These soilless systems can be classified in two groups: liquid (water) and solid (artificial substrates that are either inorganic or organic). Systems using water as a growing medium are the nutrient film technique (NFT), deep flow technique (DFT), and aeroponics. Common inorganic media are rockwool, vermiculite, perlite, and clay pellets. Organic substrates are peat, coconut coir, sawdust, and straw. Inorganic and organic substrates are usually contained in bags, and plants are irrigated with a complete nutrient solution distributed by a drip irrigation system. The excess of nutrient solution can either be allowed to leak into the ground or is recuperated and recirculated (after treatment) to plants. In liquid systems, plant roots are continuously exposed to nutrient solution, which is not leaked into the ground.

Growing methods using artificial substrates or water are known as soilless culture or hydroponics. Hydroponics is literally defined as the growing of plants in water, but the plants are actually grown in a complete nutrient solution. Ideally, the term *hydroponics* should be reserved for water culture, and the term *soilless culture* for plant cultivation on artificial substrates. In practice, the terms *hydroponics* and *soilless culture* are used indiscriminately to describe water and substrate-based systems.

Although official statistics are unavailable, hydroponic systems are known to be used extensively for food production in greenhouses. The most popular soilless medium for hydroponic vegetable production is rockwool. The nutrient film technique is also often used, but to a much lesser extend than rockwool. In some regions, the availability of low-cost materials may provide alternative substrates. For example, in British Columbia, sawdust, a residue of the large forestry industry, is commonly used as a substrate. Both aeroponics and DFT remain in little use today.

FOOD QUALITY AND SAFETY

Organic food is widely believed by the lay public to be healthier than conventional food, although the research is inconclusive. Animals fed organic diets appear to have slightly better health and reproductive performance, but similar tests in humans have not been performed. In some vegetables and cereals there is a lower concentration of protein, but it is of higher-quality. Nutrients appear to be similar with the exception of a trend towards slightly higher vitamin C in organic food. Only tentative conclusions can be drawn on the relative safety of organic food. Organic produce is likely to have less agrochemical residues, but these residues are generally below the acceptable daily intake and their health impact is questionable. Organic food also appears to have lower nitrate concentrations, but the health impact of nitrates is debated. Both organic and conventional food are expected to have similar concentrations of persistent organic pollutants and heavy metals. Data is limited on natural plant pesticides and their health effects, as well as the relative risks from bacterial pathogens.

Concerns have been raised that the higher expense of organic food (ranging from 45 to 200 per cent) could limit the recommended consumption of 5 servings per day of vegetables and fruits, which are known to improve health and reduce cancer regardless of whether they are organic or conventional. Two studies have found that children fed organic diets experienced significantly lower organophosphorus pesticide exposure than children fed conventional diets. Although the researchers did not collect health *outcome* data in this study, they concluded "it is intuitive to assume that children whose diets consist of organic food items would have a lower probability of neurologic health risks". A 2007 study found that consumption of organic milk is associated with a decrease in risk for eczema, although no comparable benefit was found for organic fruits, vegetables, or meat.

Extensive scientific research is being carried out in Switzerland at over 200 farms to determine differences in the quality of organic food products compared to conventional in addition to other tests. The FiBL Institute has been investigating the differences at over 200 farms. It states that "organic products stand out as having higher levels of secondary plant compounds and vitamin C. In the case of milk and meat, the fatty acid profile is often better from a nutritional point of view.

As far as carbohydrates and minerals, organic products are no different from conventional products. However, in regard to undesirables such as nitrate and pesticide residues, organic products have a clear advantage. A £12m EU-funded investigation into the difference between organic and ordinary farming published in 2007 found that organic foods have more nutritional value. A recent study found that organically grown produce has double the flavonoids, an important antioxidant.. A 2007 study found that organically grown kiwifruit had more antioxidants than conventional kiwifruit.

Clothing Quality and Safety

Recently, organic clothing has become widely available, due to both ecological concerns and personal health concerns. Although many consumers of organic clothing merely have a dislike of synthetic chemicals, a significant portion of the organic clothing market comes from those suffering from Multiple Chemical Sensitivity, a chronic medical condition characterized by symptoms that the affected person says are adverse effects from exposure to low levels of chemicals. Ecological concerns primary focus around pesticide use, as 16 per cent of the world's pesticides are used in the product of cotton.

Genetically Modified Organisms

A key characteristic of organic farming is rejection of genetically engineered products, including plants and animals. On October 19, 1998, participants at IFOAM's 12th Scientific Conference issued the Mar del Plata Declaration, where more than 600 delegates from over 60 countries voted unanimously to exclude the use of genetically modified organisms in food production and agriculture. From this point, it became widely recognized that GMOs are categorically excluded from organic farming.

Although opposition to the use of any transgenic technologies in organic farming is strong, agricultural researchers Luis Herrera-Estrella and Ariel Alvarez-Morales continue to advocate integration of transgenic technologies into organic farming as the optimal means to sustainable agriculture, particularly in the developing world. Similarly, some organic farmers question the rationale behind the ban on the use of genetically engineered seed because they see it a biological technology consistent with organic principles

Although GMOs are excluded from use in organic farming, there is concern that the pollen from genetically modified crops is increasingly contaminating organic and heirloom genetics making it difficult, if not impossible, to keep these genetics from entering the organic food supply. International trade restrictions limit the availability GMOs to certain countries. The actual dangers that genetic modification could pose to the environment or, supposedly, individual health, are hotly contended.

Soil Conservation

In *Dirt: The Erosion of Civilizations*, geomorphologist David Montgomery outlines a coming crisis from soil erosion. Agriculture relies on roughly one meter of topsoil, and that is being depleted ten times faster than it is being replaced. No-till farming, which some claim depends upon pesticides, is regarded as one way to minimize erosion. However, a recent study by the USDA's Agricultural Research Service has found that manure applications in organic farming are better at building up the soil than no-till despite tillage.

Climate Change

In *The Organic Answer to Climate Change*, Anthony Meleca argues that organic agriculture — with its emphasis on closed nutrient cycles, biodiversity, and effective soil management — has the capacity to mitigate and even reverse the effects of climate change. According to the Rodale Institute, which has been comparing organic agricultural systems and conventional systems since 1981, organic agriculture also can be used to mitigate global warming by decreasing fossil fuel emissions and sequestering carbon in the soil. The elimination of synthetic nitrogen in organic systems decreases fossil fuel consumption by 33 percent (LaSalle) and carbon sequestration takes CO_2 out of the atmosphere by putting it in the soil in the form of organic matter which is often lost in conventionally managed soils. Carbon sequestration occurs at especially high levels in organic no-till managed soil according to the Rodale Institute.

Nutrient Leaching

Excess nutrients in lakes, rivers, and groundwater can cause algal blooms, eutrophication, and subsequent dead zones. In addition, nitrates are harmful to aquatic organisms by themselves. The main contributor to this pollution is nitrate fertilizers whose use is expected to "double or almost triple by 2050". Researchers at the United States National Academy of Sciences found that organically fertilizing fields "significantly harmful nitrate leaching" over conventionally fertilized fields: "annual nitrate leaching was 4.4-5.6 times higher in conventional plots than organic plots".

Scientists believe that the large dead zone in the Gulf of Mexico is caused in large part by agricultural pollution: a combination of fertilizer run-off and livestock manure run-off. A study by the United States Geological Survey (USGS) found that over half of the nitrogen released into the Gulf comes from agriculture.

The economic cost of this for fishermen may be large, as they must travel far from the coast to find fish. At the 2000 IFOAM Conference, researchers presented a study of nitrogen leaching into the Danube River. They found that nitrogen run-off was substantially lower among organic farms and suggested that the external cost could be internalized by charging 1 euro per kg of nitrogen released. A 2005 study found a strong link between agricultural run-off and algae blooms in California.

Biodiversity

A wide range of organisms benefit from organic farming, but it is unclear whether organic methods confer greater benefits than integrated agri-environmental conventional programmes. Nearly all non-crop, naturally-occurring species observed in comparative farm land practice studies show a preference in organic farming both by population and richness.

Spanning all associated species, there is an average of 30 per cent more on organic farms versus conventional farming methods. Birds, butterflies, soil microbes, beetles, earthworms, spiders, vegetation, and mammals are particularly affected. Organic crops use little or no herbicides and pesticides and thus biodiversity fitness and population density benefit. Many weed species attract beneficial insects that improve soil qualities and forage on weed pests. Soil-bound organisms often benefit because of increased bacteria populations due to natural fertilizer spread such as manure, while experiencing reduced intake of herbicides and pesticides commonly associated with conventional farming methods.

Increased biodiversity, especially from soil microbes such as mycorhizzae, have been proposed as an explanation for the high yields experienced by some organic plots, especially in light of the differences seen in a 21-year comparison of organic and control fields. The level of biodiversity that can be yielded from organic farming provides a natural capital to humans. Species found in most organic farms provides a means of agricultural sustainability by reducing amount of human input (*e.g.* fertilizers, pesticides).

Farmers that produce with organic methods reduce risk of poor yields by promoting biodiversity. Common game birds such as the ring-necked pheasant and the northern bobwhite often reside in agriculture landscapes, and are a natural capital yielded from high demands of recreational hunting. Because bird species richness and population are typically higher on organic farm systems, promoting biodiversity can be seen as logical and economical. Biological research on soil and soil organisms has proven beneficial to the system of organic farming.

Varieties of bacteria and fungi break down chemicals, plant matter and animal waste into productive soil nutrients. In turn, the producer benefits by healthier yields and more arable soil for future crops. Furthermore, a 21-year study was conducted testing the effects of organic soil matter and its relationship to soil quality and yield. Controls included actively managed soil with varying levels of manure, compared to a plot with no manure input. After the study commenced, there was significantly lower yields on the control plot when compared to the fields with manure. The concluded reason was an increased soil microbe community in the manure fields, providing a healthier, more arable soil system.

Sales and Marketing

Organic farmers report that marketing and distribution are difficult obstacles. Most of organic sales are concentrated in developed nations. These products are what economists call credence goods in that they rely on uncertain certification.

As food prices rise, organic products may experience a decrease in quantity demanded. A 2008 survey by WSL Strategic Retail found that interest in organic products had dropped since 2006, and that 42 per cent of Americans polled don't trust organic produce. The Hartman Group reports that 69 per cent of Americans claim to occasionally buy organic products, down from 73 per cent in 2005. The Hartman Group says that people may be substituting local produce for organic produce.

Distributors

In the United States, 75 per cent of organic farms are smaller than 2.5 hectares and in California 2 per cent of the farms account for over half of the sales. Groups of small farms join together in cooperatives such as Organic Valley, Inc. to market their goods more effectively. Over the past twenty years, however, most of these cooperative distributors have merged or been bought out. Rural sociologist Philip H. Howard has researched the structure and transformation of the organic industry in the United States. He claims that in 1982 there were 28 consumer cooperative distributors but as of 2007 there are only 3, and he has created a graphic displaying the consolidation. His research shows that most of these small cooperatives have been absorbed into large multinational corporations such as General Mills, Heinz, ConAgra, Kellogg, and assorted other brands. This consolidation has raised concerns among consumers and journalists of potential fraud and degradation in standards. Most of these large corporations sell their organic products through subsidiaries, allowing them to keep their names off the labels.

Capacity Building

Organic agriculture can contribute to meaningful socioeconomic and ecologically sustainable development, especially in poorer countries. On one hand, this is due to the application of organic principles, which means efficient management of local resources and therefore cost-effectiveness. On the other hand, the market for organic products – at local and international level – has tremendous growth prospects and offers creative producers and exporters in the South excellent opportunities to improve their income and living conditions.Organic Agriculture is a very knowledge intensive production system. Therefore capacity building efforts play a central role in this regard. There are many efforts all around the world regarding the development of training material and the organization of training courses related to Organic Agriculture. Big parts of existing knowledge is still scattered and not easily accessible. Especially in Developing Countries this situation remains an important constraint for the growth of the organic sector.

For that reason, the International Federation of Organic Agriculture Movements created an Internet Training Platform whose objective is to become the global reference point for Organic Agriculture training through free access to high quality training materials and training programmes on Organic Agriculture. In November 2007, the Training Platform hosted more than 170 free manuals and 75 training opportunities.

Controversy

A number of critics contest the notion that organic agricultural systems are more friendly to the environment and more sustainable than high-yielding farming systems. Among these critics are Norman Borlaug, father of the "green revolution," Nobel Peace Prize laureate, who asserts that organic farming practices can at most feed 4 billion people, after expanding cropland dramatically and destroying ecosystems in the process, and Prof A. Trewavas. The debate has been summarized in an exchange between Trewavas and Lord P. Melchett, and published by a major supermarket, concerned about examining the issues. One study from the Danish Environmental Protection Agency found that, area-for-area, organic farms of potatoes, sugar beet and seed grass produce as little as half the output of conventional farming.

In 2008 a study from UN Environmental Programme concluded that organic methods greatly increase yields in Africa and a review of over two hundred crop comparisons argued that organic farming could produce enough food per capita to sustain the current human population; the difference in yields between organic and non-organic methods were small, with non-organic methods resulting in slightly higher yields in developed areas and organic methods resulting in slightly higher yields in developing areas. That analysis has been severely criticised by Alex Avery, who contends that the review claimed many non-organic studies to be organic, misreported organic yields, made false comparisons between yields of organic and non-organic studies which were not comparable, counted high organic yields several times by citing different papers which referenced the same data, and gave equal weight to studies from sources which were not impartial and rigorous university studies.

Urs Niggli, director of the FiBL Institute contends that the wave of newspaper articles like 'Organic food exposed' or 'The hypocrisy of organic farmers' are a part of a global campaign against organic farming that take their arguments mostly from the book 'The truth about organic farming', by Alex Avery of the Hudson Institute. In 1998, Dennis Avery of the Hudson Institute claimed the risk of E. coli infection was eight times higher when eating organic food rather than non-organic food, using the Centre for Disease Control (CDC) as a source. When the CDC was contacted, it stated that there was no evidence for the claim. The *New York Times* commented on Avery's attacks: "The attack on organic food by a well-financed research organization suggests that, though organic food accounts for only 1 percent of food sales in the United States, the conventional food industry is worried." Organic agriculture can reduce the level of negative externalities from (conventional) agriculture. Whether this is seen as private or public benefits depends upon the initial specification of property rights. However, it is clear that agriculture has been undervalued and underestimated as a means to combat global climate change. Soil carbon data recorded by The Rodale Institute show that regenerative organic agricultural practices are among the most effective strategies for mitigating CO_2 emissions.

8

Transpiration

A gentle breeze is relatively much more effective in increasing the transpiration rate than winds of greater velocity. Winds of very high velocity have been observed to have a retarding effect upon transpiration. This is probably due to closure of stomates under such conditions.The effect of wind upon the transpiration of a leaf which is exposed to intense insolation is further complicated by the fact that a wind may have a cooling effect upon the leaves, due to their increased thermal emissivity under such conditions.

Reduction in the temperature of a leaf, as we have already seen, exerts a decelerating effect upon transpiration. If conditions are such that this cooling effect of wind is predominant increased wind velocity may result in a diminution in transpiration rate.

SOIL CONDITIONS INFLUENCING THE AVAILABILITY OF WATER

Although transpiration can continue for short periods at rates considerably in excess of the rate of absorption of water, in general, if soil conditions are such that absorption of water is appreciably retarded, the rate of transpiration will soon show a corresponding retardation.

The availability of soil water to the plant is therefore an important and, in fact, often the limiting factor in transpiration.

The principal soil factors which affect the rate of absorption of water by plants are:

- Available soil water,
- Soil temperature,
- Aeration of the soil, and
- Concentration of solutes in the soil solution. All of these factors indirectly influence the rate of transpiration.

FACTORS AFFECTING TRANSPIRATION

ATMOSPHERIC PRESSURE

It has been demonstrated experimentally that a reduction in atmospheric pressure results in an increase in the rate of transpiration. This result would be predicted theoretically since reduction in the density of the atmosphere would be expected to permit diffusion of water-vapour to occur into it more rapidly. In any given locality variations in atmospheric pressure are too slight to have any significant effect upon the rates of transpiration. Plants growing at high altitudes are subjected to distinctly lower atmospheric pressures thanthe plants of lowlands, and in comparative evaluations of the transpiration rates of species growing in these two types of habitats the influence of this factor must be considered.

STRUCTURAL FEATURES OF PLANTS WHICH INFLUENCE THE RATE OF TRANSPIRATION

A number of diverse species of plants growing under identical environmental conditions usually differ very greatly in transpiration rates. Physiological factors, such as differences in stomatal behaviour, cell sap concentration, colloid content of the cells, etc. are undoubtedly partially responsible for the fact that all plants do not lose water-vapour at the same rate even under the same environmental conditions. Structural differences, particularly in the leaves, also account in part for unlike rates of transpiration of different species growing in the same environment.

Many of the supposed effectsof structural differences in plants upon their transpiration rates are, however based solely on inferences, and not upon experimentally determined facts. Many species which on the basis of their anatomy have been judged to have a low transpiration rate later were actually found to transpire very rapidly when environmental conditions were suitable. The following discussion of the anatomical features known or generally believed to influence the rate of transpiration will therefore be very brief.

THICKNESS OF THE CUTIN

It can be easily demonstrated that water vapourates more readily from uncutinized epidermal cell walls than from those which are coated with a layer of cutin. It is sometimes assumed, therefore that the thicker the layer of cutin on the outer walls of the epidermis of a leaf, the slower the rate of cuticular transpiration from that leaf. Actually,however, it seems obvious that any increased thickness in the cutin layer beyond that at which cuticular transpiration becomes negligible will have no appreciable influence on the rate of evapouration from the epidermal cell walls.

EPIDERINAL HAIRS

H airs are prominent features of one or both of the surfaces of the leaves of many species. Living hairs probably increase the rate of cuticular transpiration by increasing the exposed surface of the leaf. Many botanists have assumed that dead hairs reduce the rate of stomatal transpiration, particularly under conditions of intense sunlight or strong winds. In the former case it has been supposed that the whitish hairs reflect such a large proportion of the incident light that the leaf temperature is not as high as it otherwise would be, and that the rate of transpiration is correspondingly lower. In the latter case it has been supposed that the hairs impose a mechanical barrier to the effect of wind on transpiration. However, Sayre found that removal of the hairs from the leaves of the mullein had little effect upon the rate of stomatal transpiration under ordinary conditions of light intensity and wind velocity, although such treatment did result in an increase in cuticular transpiration.

RATIO OF INTERNAL TO EXTERNAL SURFACE IN LEAVES

Although transpiration rates are most frequently expressed in terms of the exposed area of the leaf surface, most of the water-vapour lost from leaves evapourates from the walls of the mesophyll cells which bound the intercellular spaces. The area of this internal evapourating surface in proportion to the external surface of a leaf varies greatly not only in leaves of one species as compared to those of another, but in leaves of the same species if they have developed under different environmental conditions.

Computations of the ratio of internal exposed surface to external exposed surface of leaves of a number of species have yielded values ranging from 6.8 to 31.3. In general the greater the proportion of palisade to sponge tissue the greater this ratio.The thickness of the leaf will also obviously be a factor influencing this ratio.The ratio is greater in sun leaves than in shade leaves of the same species; for the lilac the values are 13.2 and 6.8, respectively.

It has generally been supposed that the transpiration rates of leaves in which a relatively large area of mesophyll cell walls is exposed to the intercellular spaces as compared with the exposed epidermal area of the leaf are greater than in leaves in which the contrary condition obtains. Theoretically such an effect could be exerted only if the area of exposed mesophyll cell walls exercised a controlling influence upon the steepness of the vapour pressure gradient through the stomates.

When transpiration is relatively high a steeper vapour pressure gradient may be maintained by leaves possessing relatively extensive internal evapourating surfaces than by those which do not. The higher transpiration rate of sun leaves as compared with shade leaves on the same plant, which has been observed in some species, may be partly explainable on this basis.

SUNKEN STOMATES

The stomates of many species of plants are sunken below the general level of the epidermis. Since the length of a diffusion gradient is one of the factors governing its steepness, diffusion through a small opening is slower if the gas must pass through a relatively long tube before reaching the orifice, than if the diffusion is through a shorter tube. In leaves in which the stomates are not sunken the diffusion column is relatively short, amounting only to the diameter of the guard cells as seen in cross section, while in sunken stomates the column of gas may be longer by several times.

The diffusion of gases through such stomates is undoubtedly slower than through stomates of the ordinary type. This retarding effect of sunken stomates upon the rate of transpiration as compared with the stomates of the non-sunken type has been estimated by Renner to range from 30 to 70 per cent.

STRUCTURE AND DISTRIBUTION OF THE ROOT SYSTEM

Under some environmental conditions an individual plant of a given species will develop a greater leaf surface in proportion to the extent of its root system thanunder others. The latter type of structural development will, other conditions being equal, favour the maintenance of higher transpiration rates thanthe former.

The distribution of a root system in the soil sometimes influences rates of water absorption and transpiration. In a habitat where deep-rooted and shallow-rooted species are growing side by side the former may transpire more rapidly during drought periods than the latter because their roots penetrate to a soil horizon which still contains available water, while roots of the latter are in a moisture deficient soil. The completeness with which the soil mass is interpenetrated by roots may also influence rates of absorption and transpiration.Thus Miller found that sorghum plants have almost twice as many fibrous roots as maize plants. Rates of absorption and transpiration can be better maintained, especially when the soil is relatively dry, by plants with the sorghum type of root system than by plants with the maize type of root system.

THE DAILY PERIODICITY OF TRANSPIRATION

All plants exhibit a daily periodicity of transpiration rate which varies somewhat with the species and is greatly influenced by the environmental conditions to which the plant is exposed. We shall first consider the transpiration periodicity upon a standard day as defined in the preceding chapter, since most experiments upon daily variations in transpiration rate have been conducted upon plants growing under approx-imately such conditions.

It should not be assumed, however, that under natural conditions the daily periodicity of transpiration usually or normally follows the trend. The daily periodicity of transpiration under these roughly defined conditions does, however, afford a convenient reference standard with which to compare the many variations which are manifested by this phenomenon.

The transpiration rate during the hours of darkness is generally low, and in most species water loss during this period may be regarded as almost entirely cuticular. It is not justifiable to assume that absolutely no stomatal transpiration occurs at night in any species unless this is definitely known to be true. In some species of plants, as we have already seen, nightstomatal opening is a common occurrence, while in others it occurs under certain environmental conditions. Even in those species in which the stomates are normally closed during the hours of darkness, a few may remain open.

The transpiration rate shows a rapid rise during the early morning hours followed by a slower but consistent increase which culminates in a point of maximum transpiration which occurs sometime during the mid-day hours. Following this peak of transpiration, the rate decreases usually consistently although often with an accelerated rate in the late after noon, until the low and virtually steady night rate is attained at approximately the termination of the daylight period.

The daily periodicity in the rate of stomatal transpiration in most species of plants can be interpreted almost entirely in terms of two variable factors:

1. The diffusive capacity of the stomates, and
2. The vapour pressure gradient between the intercellular spaces of the leaf and the outside atmosphere. In analyzing the daily periodicity of transpiration it is the diffusive capacity of the stomates in the aggregate (*i.e.* all of the stomates on a plant) which must be considered. The aggregate diffusive capacity of the stomates on a plant on which all of the stomates were half open, for example, would be much greater than that for a similar plant on which half of the stomates were fully open and half fully closed.

With the advent of daylight the stomates open over a period of time which varies in length according to the species and with environmental conditions. Some open earlier or more rapidly than others. The initial rise in the rate of transpiration in the morning is due to the opening of the stomates, resulting in a gradual increase in their aggregate diffusive capacity. As each stomate opens a vapour pressure gradient is established through it between the atmosphere of the intercellular spaces and the outside atmosphere. During the hours of darkness the leaf cells regain most or all of the water which they have lost during the preceding day, and thus attain approximately their normal turgidity.

This facilitates the evapouration of water into the intercellular spaces, so that when the stomates open in the morning be internal atmosphere of the leaf is usually saturated with water-vapour, or practically so. The vapour pressure of the intercellular spaces is therefore at the maximum possible for that leaf temperature. This is, on clear days, almost always in excess of the vapour pressure of the atmosphere when the stomates open, and often considerably so. Hence outward diffusion of water-vapour through the stomates usually begins as soon as they are open.

Once a stomate is well open, minor variations in the size of the stomatal aperture have little or no effect on the rate of water-vapour loss through it.The rate of transpiration continues to increase, however, for some time afterthe stomates, in the aggregate, have attained their maximum diffusive capacity. This is due to a gradual increase in the steepness of the vapour pressure gradient through the stomates. As the day progresses the temperature of both the atmosphere and the leaf increases; if the latter is in direct sunlight its temperature is invariably somewhat in excess of that of the atmosphere. On clear days, as previously described, the vapour pressure of the intercellular spaces usually increases relative to that of the atmosphere with a rise in temperature.

The steepness of the vapour pressure gradient between the internal atmosphere of the leaf and the external atmosphere therefore usually increases progressively during the earlier part of the day, and this is the important factor accounting for a rise in the transpiration rate once the stomates have attained approximately their maximum diffusive capacity. Almost from the moment when the stomates begin to open in the morning a train of events is set in operation in the plant which ultimately causes a reduction in the rate of transpiration.

During approximately the first half of the day, however, these factors are more than offset by the factors resulting in an increase in the rate of transpiration. In most plants, while transpiration is occurring rapidly, the rate of absorption of water does not keep pace with the rate at which water-vapour is lost from the leaves. This results in a reduction in the water content of the entire plant, and especially that of the leaves. In more extreme cases wilting results, but under standard day conditions the leaves seldom pass beyond the stage of incipient wilting which corresponds only to a partial loss of turgor by the leaf cells.

THE DAILY PERIODICITY OF TRANSPIRATION

Decrease in the water content of the leaf cells also results in a rise in their diffusion pressure deficit, and a correlated decrease in their vapour pressure. A fully turgid cell (*i.e.* one with a diffusion pressure deficit of zero) has an equilibrium vapour pressure equivalent to that of pure water regardless of its osmotic pressure.At 20 °C. this is 17.54 mm. Hg. If the diffusion pressure deficit of the mesophyll cells of a leaf at this temperature were raised to 100 atmos., a value which can be attained in only a very few species of plants, their vapour pressure would decrease by only about 1.27 mm. Hg. Since the vapour pressure of the cell walls, with which the vapour pressure of the intercellular spaces tends to be in equilibrium, in turn tends to be in equilibrium with that of the cell sap, it would appear that an increase in the diffusion pressure deficit of the mesophyll cells has only a very slight influence upon vapour pressure conditions in the intercellular spaces.

However, certain other factors must be taken into consideration in evaluating the effect of a diminished leaf water content upon the vapour pressure of the intercellular spaces. Decrease in the water content of the leaf cells involves a reduction in the quantity of water in the cell walls as well as in other parts of the cell. Water undoubtedly passes through cell walls principally through the intermicellar material. The principal effect of a decrease in the quantity of water in the walls is a shrinkage in the volume of the hydrophilic intermicellar material and a resulting contraction in diameter of the intermicellar capillaries.

This probably results in a decrease in the permeability of the walls to water. Hence, under such conditions, water often may not pass across the walls with sufficient rapidity that an equilibrium vapour pressure is maintained between the cell sap and the outer surface of the cell walls. When the water content of a leaf is low the vapour pressure of the cell wall surface in contact with the intercellular spaces may therefore be less than the vapour pressure of the water in the cell sap. Any such reduction in the vapour pressure of the intercellular spaces results in decreasing the steepness of the vapour pressure gradient between the intercellular spaces and the outside atmosphere, and hence decreases the rate of transpiration.

The experimental results of That indicate that the vapour pressure of the intercellular spaces is often at less than a saturation value. During the course of a day, however, the steepness of the vapour pressure gradient through the stomates may be decreased in still another and perhaps more important way. When the stomates open in the morning the vapour pressure gradient between the saturated internal leaf atmosphere and the outside atmosphere is short and hence steep, being approximately equal in length to the depth of the guard cells. For some time after the stomates open essentially such a condition is probably maintained. As the day advances the rapid loss of water-vapour molecules out of the interc-ellular spaces, perhaps coupled with a gradual reduction in the vapour pressure of the cell walls makes it less and less likely that a condition of saturation can be maintained throughout the intercellular spaces. The zone in which a vapour pressure inequilibrium with the water in the mesophyll cell walls is maintained almostcertainly retreats more and more deeply into the intercellular spaces. Eventually it may be restricted to a thin layer just above the evapourating surfaces of the mesophyll cell walls. This gradual lengthening of the vapour pressuregradient through the stomates is probably an important factor in bringing about a reduction in the rate of transpiration during the afternoon hours.

It has frequently been observed in studies of transpiration periodicity that the transpiration rate often begins to decrease during the midday period before any appreciable change occurs in the area of the stomatal apertures. This indicates that the initial reduction in transpiration rate is induced by some other factor than a change in the diffusive capacity of the stomates. As the preceding discussion has shown this other factor is almost certainly a reduction in the steepness of the vapour pressure gradient through the stomates.By late afternoon the air temperature and the intensity of the solar radiation begin to decrease appreciably, thus inducing a decrease in the temperature of the leaf. This lowering of the leaf temperature may further decrease the vapour pressure of the intercellular spaces, and hence further depress the steepness of the vapour pressure gradient, since temperature changes have very little influence on the vapour pressure of the outside atmosphere under standard day conditions.

Thus the effect of a reduction in the vapour pressureg radient through the stomates in decreasing the rate of transpiration, which first becomes apparent during the midday period, continues with augmented effect as the hours of darkness approach. Continued diminution in the leaf water content due to the excess of transpiration over absorption eventually results also in a diminution of the turgor of the guard cells due to a decrease in the water content of the leaf.

This results in a gradual closure of the stomates. Some of the stomates on a plant probably begin to close even before the time at which the peak of the transpiration rate is attained. In all likelihood stomates near the margin or tip of a leaf begin to close before those in the middle since the effects of deficiency of water usually appear first in these regions of a leaf. With an increasing leaf water deficit more and more of the stomates on the plant close.

This soon results in a reduction in the aggregated iffusive capacity of the stomatal population of the plant. As the afternoon advances more and more of the stomates become completely closed, resulting in a progressive diminution in diffusive capacity. This gradual closing of the stomates, resulting in an increasingly larger proportion of them being closed as the day advances, is the principal factor reducing the rate of transpiration during the latter part of the day. The effect of this factor and that of a decreased steepness of the vapour pressure gradient overlap during a large part of the afternoon. Finally, by late afternoon, complete closure of essentially all of the stomates has occurred. Stomatal transpiration is terminated at that time and during the hours of darkness the rate of water loss from the plant is controlled by factors influencing the rate of cuticular transpiration. Under environmental conditions deviating very greatly from those which were postulated in the preceding discussion, transpiration periodicity curves may be entirely different from those. Variations in temperature, light intensity, humidity, and soil water supply may all markedly influence both the trend of transpiration periodicity, and the magnitude of the daily water loss.

Low temperatures may result in a complete elimination of stomatal transpiration by inducing stomatal closure. Low light intensities, such as those existing on cloudy days, are unfavourable to stomatal opening in most species. The stomates seldom open completely under such conditions and the period during which they are open is usually of shorter duration than on clear days. Furthermore, the vapour pressuregradient through the stomates is seldom as steep on such days as on clear bright days, since leaf temperatures never appreciably exceed atmospheric temperatures except when exposed to direct sunlight, and atmospheric vapour pressures are usually higher during cloudy days than on clear days at the same season of the year. The magnitude of transpiration under such conditions is usually greatly reduced, and transpiration periodicity curves plotted for plants exposed to such conditions usually present a somewhat fore shortened and greatly flattened appearance.

A deficient soil water supply is most commonly the factor which causes marked departures from the type of transpiration periodicity already considered, especially during the summer months. The effect of a gradual diminution in soil water content upon transpiration periodicity.

A reduction in soil water content has two pronounced effects upon the daily march of transpiration. The total daily magnitude of water loss is decreased, and the peak of the transpiration curve often occurs somewhat earlier in the day than under conditions of abundant soil water supply. Since, even in temperate regions, periods of decreased soil water supply are of common occurrence during the summer months, and in many habitats are the rule rather than the exception for many species of plants much of the time.

Internal factors may also be responsible for transpiration periodicities of a different trend from those which have already been discussed. In some species of plants, as has already been described, the stomates remain open to a greater or lesser extent during the hours of darkness. Such plants have higher transpiration rates at night than those in which the stomates are closed. In some species of cacti there is a complete inversion of the usual transpiration periodicity curve, transpiration rates being regularly greater at night than in the daytime. This appears to be due to complete or nearly complete stomatal closure during the hours of daylight, while the stomates are, as a rule, open at night.

SEASONAL VARIATIONS IN TRANSPIRATION RATES

In temperate regions transpiration occurs predominantly during the warmer months of the year and especially during those periods of the warm season when the soil water supply is abundant. Among the deciduous group of woody perennials, the branches are defoliated during most of the autumn, all of the winter, and the earlier part of the spring.

Although the twigs and branches which remain exposed to the atmosphere are completely encased in corky layers of bark and the buds are enclosed within cutinized bud scales some water loss occurs from such species even during the winter months. Young twigs lose water under such conditions faster than older ones. Winter transpiration rates of deciduous woody plants are always negligible in comparison with summer rates. In evergreen species of either the needle-leaved or broad-leaved type the transpiration rates are usually not appreciably different at most times during the winter from the transpiration rate of deciduous trees at that season.This is probably due principally to the fact that the low temperatures of the winter prevent opening of the stomates although the relatively low leaf water contents characteristic of this season may also be a factor in maintaining the stomates in a closed condition. Mild periods of any considerable duration often have a detrimental effect upon evergreen species during the winter months. The warm air temperatures induce stomatal opening, and a relatively high rate of transpiration ensues.

The soil is apt to be deficient in water at such times, or even if present, that in the surface layers may remain frozen during a period of warm air temperatures. Even if neither of the foregoing conditions exists, the soil temperatures will be low, and this greatly retards the rate of absorption of water.

The combined effect of a relatively high transpiration rate and relatively low absorption rate results in a gradual desiccation of the leaves and branches of the plant. This diminution in the water content of the aerial organs during warm periods in winter is more severe in windy weather and is more likely to occur if a sequence of mild days follows immediately after a cold spell. If severe enough this desiccation will result in the death of some of the branches, or in extreme cases of the entire tree or shrub. This is one cause of the phenomenon known as winter-killing. Winter-killing due to desiccation of the tissues should not be confused with cold injury due to low temperatures which is an entirely different phenomenon.

THE MOVEMENT OF WATER THROUGH THE PLANT

Most terrestrial plants obtain the water which is necessary for their existence from the soil. An overwhelmingly large proportion of the water which is absorbed by the roots of land plants is lost in the process of transpiration. Smaller quantities are utilized in growth and in photosynthesis and in some species limited amounts of water may be lost by guttation.

Water must therefore move through the intervening tissues and organs from the absorbing regions of the root to the tissues in which it is utilized, or from which it passes out of the plant. The process whereby water moves through the plant is often termed the conduction, transport, or translocation of water.

In herbaceous species and many shrubby plants the distance throughwhich water moves in passing from the root tips to the leaves is usually not more than a few feet. Even in such plants appearances are sometimes deceptive, as some herbaceous and shrubby species such as alfalfa may have such deep root systems that some of the absorbed water often ascends for distances as great as twenty or more feet before it reaches the level of the soil surface. It is in trees, however, in which the most striking illustrations of the upward movement of water occur. The tallest tree of which we have an authentic record is a specimen of the Coast Redwood which has attained a height of 364 feet. Many other individuals of this species and of several others, including the Big Trees of California, the Douglas Firs (Pseudotsuga mucronata) of the Pacific Northwest, and the Blue Gums (Eucalyptus) of Australia exceed 300 feet in height. Trees ranging from 100 to 200 feet in height were of common occurrence in the virgin forests of eastern North America.

Since the root systems of trees always penetrate at least a few feet into the ground the actual vertical distance through which at least a part of the water absorbed is elevated is always more than the height of the tree. In trees, therefore, water must ascend to heights ranging up to nearly 400 feet above the level of water absorption.

The mechanism by which this feat is accomplished in tall trees has been the subject of much experimentation, and even more speculation. The ensuing discussion of the ascent of sap through plants will be presented largely in terms of its movement through trees, principally because much of the experimental work on this problem has been performed on woody species. Any explanation of this phenomenon which can be shown to be adequate for tall trees should also prove satisfactory for vascular species of lesser stature.

THE PATH OF WATER THROUGH THE PLANT

Water enters the plant through the epidermal cells and root hairs at or near the tips of the roots and crosses the cortex, endodermis, and a part of the pericycle before it finally enters the lumina of the vessels or tracheids of the root xylem. At this point its upward movement begins. The xylem tissue is continuous from just back of the root tips, through the roots, into and through the stems, the petioles of leaves, and ultimately, usually only after much branching, terminates in the mesophyll of the leaf.

The xylem tissue through which the water moves is thus a continuous unit system within the body of the plant. Along most of its course water moves en masse through the vessels or tracheids of the xylem. At the termination of the xylem ducts in the leaves the water passes into the adjacent mesophyll cells.

In the mesophyll of the leaves it moves from cell to cell, eventually most of it being lost from the cells by ognized at least since the time of the girdling experiments of Malpighi in 1671. Girdling (or ringing) a stem, so that all of the tissues external tothe xylem are removed, does not prevent movement of water to organs attached to that stem above the ring. On the contrary cutting through the xylem tissue of a stem results in almost immediate wilting of leaves attached to the stem above the ring.

ANATOMY OF STEMS

Since the mechanism of the movement of water can scarcely be understood intelligently without some knowledge of the anatomy of the tissues through which it moves, a brief review of stem structure is desirable before proceeding further with a discussion of this process.

Stems vary greatly in their structure, every species possessing some anatomical features which are peculiar to itself. Nevertheless certain general patterns of tissue arrangement have been found to prevail, and the stem structure of most species will approximate one or another of these general arrangements.

The corn stem represents an herbaceous monocot type of structure while that of the tulip poplar is typical of woody dicot stems. The structure of woody stems cannot properly be appreciated merely from a consideration of one year old stems.

The primary tissues of such stems develop from tissues formed at the apical growing tip during growth of the stem in length. Nearly all perennial stems also grow in diameter as a result of the development of secondary tissues from the cambium.

By successive stages of division, enlargement, and differentiation of cambium cells additional layers of xylem (secondary xylem) are laid down on the inner face of the cambium, and new layers of phloem tissue (secondary plzloem) on its outer face Hence after a few years the great bulk of any woody stem is composed of secondary tissues. Secondary xylem and secondary phloem also develop from the cambium in most herbaceous dicot stems but in such species the cambium in any one of phloem and xylem elements from the cambium.

- A cambium initial,
- Division of cambium cell,
- Outer cell resulting from division becomes a phloem element,
- Another division of the cambium cell,
- Outer cell resulting from second division becomes a xylem element.

Actually several xylem or several phloem elements are often formed successively. Upon maturation most of the xylem and phloem elements acquire sizes and shapes which are very different from those of the cambium initials from which they originate. Stem is never active for more than one growing season. The spring formed xylem tissue, as viewed in cross section is usually distinctly different in aspect from that formed later in the season.

In many angiosperms the springwood contains more and larger vessels and the cell walls are generally thinner than in the subsequently produced summer wood. In the conifers the spring formed tracheids are thinner-walled and of larger cross-sectional diameter than those formed later in the growing season. The transition from spring to summer wood is often a very gradual one. On the other hand the more open xylem tissues formed each spring abut directly upon the denser tissues which were produced during the preceding summer, thus giving rise to an abrupt line of demarcation between the zones of xylem formed in any two successive seasons.The result of this growth behaviour is that a cross section of the trunk or a branch of any tree appears as a system of concentric layers, the so-called annual rings, each representing an annual increment of growth. In rare cases no annual rings nor more than one annual ring may be produced in a season, but usually each ring represents the xylem produced by the activity of the cambium during one season.In many woody species as the xylem tissues increase in age important changes occur in the colour, composition, and structure of the various elements, resulting in the conversion of sapwood into heartwood.

As sapwood ripens into heartwood the walls of any remaining living cells of the xylem become increasingly lignified death of these cells soon following. The water content of the tissues is generally reduced and such compounds as oils, resins gums, and tannins accumulate in the cells or cell walls. The darker colouration of the heartwood of most species as compared to the sapwood is due to such accumulations. In mature trees the heartwood becomes merely a central supporting column surrounded by a cylinder of sapwood which varies in thickness from a few to many annual layers, depending upon the species and the environmental conditions under which the tree is growing.

In some species (apple, elm) the heartwood remains virtually saturated Tangential section of a small portion of the wood of whitepine (Pinus strobus). The vertically oriented elongated elements are tracheids. Bordered pits show in sectional view in the tracheid and marginal ray cells. Simple pits show in sectional view in other ray cells. Half-bordered pits show between marginal ray cells and other ray cells.

The water in the heartwood of such species as apple and elm appears to be largely static and is not directly involved in translocation. Coincident with the development of secondary xylem, secondary phloem tissues also develop from the cambium. Cork cambiums are also initiated in the bark which produce cork layers. Profound modifications therefore occur in the outer tissues as well as in the xylem of woody stems as they grow older.

A somewhat more detailed concept of the structure of the cells and elements of the stems of angiosperms through which the movement of water occurs is presented which represent longitudinal tangential and radial sections from the xylem of a tulip tree which may be taken as representative for this group of plants. The xylem tissues of the wood of angiosperms are composed of vessels tracheids, fibres of several types, wood parenchyma, and xylem ray cells. There is a tremendous variability in the proportional distribution and arrangement of these tissues according to species.

The most characteristic elements of the xylem tissue of angiosperms are the vessels. These are, in general, more or less tubular structures which may extend through many feet of the xylem. In some species cross walls, usually perforated, are of frequent occurrence in vessels, in others such cross walls are infrequent or lacking. In diameter they may range in trees from about 20 to about 400. In vines they may be as much as 700 in diameter. The vessels branch extensively in certain regions of the plant, especially at nodes within the leaf lamina, and at the points in the root systems where rool branching occurs.

The vessels of the proto xylem (the first cells of the xylem to mature during the ontogeny of a growing stem or root tip) have cellulose walls which are reinforced by distinctive lignified thickenings which appear as rings, spirals,or other characteristic patterns. The vessels that develop later in the ontogeny of any growing tip have lignified, usually pitted walls which lack any of the thickenings which distinguish the walls of the vessels of the protoxylem.Pits normally occur in all parts of vessel walls which are contiguous with other vessels or cells. These pits consist essentially of thin areas in the walls. The walls of the vast majority of plant cells are pitted. The architecturally distinct type known as the bordered pit.

FUNCTION OF STEM

Tracizeids are found in wood of many but not all species of angiosperms. They are typically more or less spindle-shaped cells with thick walls which are almost always lignified. As viewed in cross section they are usually angular. Mature tracheids contain no protoplasm and hence are non-living like the xylem vessels. The largest tracheids are about 5 mm. in length, and about 30 in diameter, although most of them are smaller. The walls of tracheids are pitted like those of vessels. Tracheids are water-conducting cells but in most angiosperms they are relatively of much less importance as channels throughwhich water moves than the vessels.

Tracheids are developed from the cambium by a process essentially similar to that by which vessel segments are formed, except that the size and shape of the resulting cells is different, and that there is no coalescence of the individual elements such as occurs in the formation of vessels. The xylem of the angiosperms also contains fibres of various types, which in general are similar to the tracheids in structure except that usually they have thicker walls, smaller lumina, and fewer and smaller pits. Fibre cells are non-living and their walls are lignified. Because of their thick walls and small lumina the fibres do not play an important part in the movement of water through stems.

The xylem tissue of practically all angiosperms also contains wood parenchyma cells which, unlike the elements already described, remain alive for some time after their differentiation. Usually death of the wood parenchyma cells does not occur until the wood of which they are a part is converted into heartwood. Wood parenchyma cells are generally somewhat elongate, and roughly four-angled in cross section. They occur in the xylem as vertical series of cells placed end to end. The strands of wood parenchyma thus produced often extend for a long distance in a vertical direction through the wood.The distribution of the wood parenchyma strands throughout an annual ring of xylem tissue varies with the species. In some they are more or less scattered through the xylem, in others they occur only in the last layer or two of cells which are produced in the summer woodin other words at the termination of the seasons growth and in others only in contact with the vessels, or in contact with other wood parenchyma cells which are themselves in contact with the vessels. All of the xylem elements previously described are oriented with their long axes in a vertical direction.

In addition to this vertical system there is also present in the xylem a transverse, radiating system in which the long axis of the cells is at right angles to the long axis of the stem. These transversely oriented tissue units are known as vascular rays. In the stems of most species they are continuous from the outer extremity of the phloem through the cambium into the xylem, which they penetrate to a greater or lesser distance. The portion of the vascular ray found in the xylem is termed the xylem ray; that found in the phloem the pliloem ray. Xylem rays may vary from one to many cells in thickness and likewise in height, certain types usually being characteristic of any one species. The cells of the xylem rays,like those of the wood parenchyma, usually remain alive until the woody tissue is converted into heartwood. The cells of the xylem ray are typically elongate and more or less angular in cross section.

The xylem rays probably serve as routes along which lateral movement of water occurs from the xylem to the cambium and phloem, and along which translocation of soluble foods takes place from the phloem to the living cells of the xylem. The living ray cells are in contact at various points with the strands of living wood parenchyma cells. The vertically oriented wood parenchyma strands and transversely oriented xylem ray strands thus form a unit system of living cells within the woody cylinder.

Hence there is present a continuous inter meshing network of living cells throughout the greater mass of non-living vessels, tracheids, and fibres in the younger portions of any woody angiosperm stem. There are probably few if any of the conducting elements vessels and tracheids which are not in contact at one or more points with this continuous system of living cells. The wood of the gymnosperms is simpler in its structure than that of the angiosperms, this group of woody species also showing, in general, a greater uniformity in stem structure than the latter group. The only cell types universally present in the wood of coniferous trees are the tracheids and woodray cells. Wood parenchyma cells are also present in the wood of most species of conifers, while in many species tracheid-like fibre cells are also present. The most important distinction between the wood of conifers and that of angiosperms is the total absence of vessels in the former. The tracheids are the distinctive element in conifer wood, constituting as they do the great bulk of all the woody tissues present in such species. In conifers the tracheids form a densely packed type of woody tissue, composed of interlocking cells. Vertically contiguous tracheids always overlap along their tapering portions. Movement of water and solutes from one tracheid to another is facilitated by means of the bordered pits in the adjacent walls. Because of the numerous cross walls in a xylem tissue composed almost entirely of tracheids, water encounters a greater resistance in moving through such tissues than in traversing woody tissues which contain vessels. Nevertheless it is interesting to note that the tallest trees in the world are conifers in which all movement of water occurs through tracheids.

Theories of the Mechanism of the Movement of Water through Plants. A number of different theories of the mechanism by which the ascent of water is brought about iii plants have been suggested, and it is probable that more than one mechanism is involved in this process.

The present state of our knowledge justifies a discussion of only three possible mechanisms:

1. That ascent of sap is caused by the vital activities of living cells, principally those in the stem,
2. That this process occurs according to a mechanism described in the cohesion of water theory, and
3. That upward movement of water occurs as a result of root pressures. In passing, it should be noted that neither atmospheric pressure nor capillarity can account for the rise of sap in plants, although both of these processes are recurringly suggested as being involved in this process. Atmospheric pressure could never account fora rise of more than about 10 metres, and most mature trees attain a greater height than this. The capillary rise of water in tubes 10 in diameter is only about 3 metres, and few if any of the conducting elements have as small a diameter as this.

The mechanism of the movement of water through the plant is inextricably bound up with the mechanism of the absorption of water, so the following discussion will necessarily treat to some extent of both of these processes. Vital Theories. While the vessels and tracheids through which longitudinal transit of water occurs are non-living, they are always in more or less intimate contact with living cells. Suggestions have therefore been made from time to time that the motive power causing movement of water is furnished by the living cells of the stem.

The most recent advocate of such a theory is Bose. As an example of the vital theories of the ascent of sap we may consider the proposals advanced by Godlewski in 1884. He considered that conduction of water occurred as a result of periodic changes in the osmotic pressure of the living cells present in the xylem (especially the wooJ ray cells). Upon increase in their osmotic pressure water was supposed to move into the cells from bordering xylem elements.

Upon decrease in osmotic pressure water was supposed to move back into a xylem duct. In the xylem duct it was supposed that water would move upward until it came in contact with another living ray cell because of a lower air pressure in the upper than in the lower part of the vessels. The ascent of water in stems was thus supposed to be brought about by a number of repetitions of this same process. There is little or no direct evidence in support of this or any of a number of more or less similar theories which have been suggested.

The experiments of Strasburger demonstrated quite clearly that the primary mechanism of the rise of sap in trees operates independently of the living cells of the stem. He performed many experiments upon the movement of water through woody stems in which the living cells had been killed by one method or another. In one experiment, for example, he used a 75 -year -old oak tree about 22 metres in height. This was sawed off close tothe ground and the cut end transferred to a solution of picric acid, which is toxic to living cells.

The picric acid solution slowly moved up the stem. Fuchsin, added to the liquid in which the base of the tree was immersed 3 days after the picric acid, also ascended to the top of the tree through tissues in which the living cells had been killed by the picric acid. Water also continued to ascend through similar stems after they had been completely killed by exposure to a temperature of 900 °C. Similar experiments have been performed by later investigators.

All of these investigators have confirmed Strasburgers results that water would continue to ascend for some time through stems, segments of which had been exposed to one treatment or another which would kill all living cells present. In all experiments of the type just described, it has been observed that the leaves at the top of a stem, part or all of which has been killed, sooner or later wilt and wither, although this effect is by no means an immediate one,often appearing only after several days. Proponents of the vital theories have accepted this as evidence that the living cells of a stem are essential for the passage of water through it. Dixon however, considers that this delayed lethal effect on the leaves is due to either or both of two causes.

Killing of the stem tissues often causes the formation of substances which plug up the vessels or tracheids, thus impeding the upward movement of water. Furthermore, death of the cells in the treated regions causes the release into the conducting channels of toxic compounds which, when transported to the eaves, cause death of the leaf cells. There is still some doubt, however,whether all of the retardation in rate of translocation as a result of such experiments can be explained in these two ways. The possibility remains open that the living cells of the xylem may in some way be necessary to the maintenance of the water-conducting capacity of the xylem, or even that they may contribute in some way to the lifting of water through stems.

In general, the evidence appears to be quite conclusive that the living cells of the stem do not furnish the principal mechanism which causes the upward transport of water in stems. On the contrary, as shown in the subsequent discussion, the living leaves or stem tips at the top of the plant appear to be absolutely necessary for the ascent of water through stems. Root Pressure. That a slow exudation of sap often occurs from the cut surface of a stem is well known. Such sap exudations are often ascribed to the development of a pressure in the dilute sap of the xylem vessels resulting from the operation of some as yet not fully understood mechanism in the root cells. Hence the name root pressure. It is generally considered that such pressures may occur in intact plants as well as in those which have been cut into. It is probable that not all of the phenomena which have been classed under the name of root pressure are due to the operation of the same physiological mechanism. Some sap exudations seem to come from the cambium or phloem rather than from the xylem. It seems quite certain that some of the sap exudations from the xylem occur from the living wood parenchyma or woodray cells rather than from the vessels, or at least that the exudation is greatly influenced by the activity of such cells.

Hence the living cells of the stemas well as those of the root are probably involved in many root pressure phenomena. James and Baker go so far as to regard the subjection of the sap in vessels to a pressure as an exceptional occurrence in plants, but this seems to be an extreme point of view, and will not be followed in this discussion. Because of the unsettled state of our knowledge regarding this phenomenon it will be necessary to use the term root pressure in a very uncritical sense.

The magnitude of root pressures is commonly determined by attaching a manometer to the cut surface of the plant from which the bleeding is occurring. With only a few rare exceptions the exudation pressures which have been recorded for plants do not exceed 2 atmos., and most of them are less than this. According to White individual excised roots of tomato plants exude sap from their basal ends with a pressure of more than 6 atmos.

It has not yet been shown, however, that such pressures operate for any considerable distance through plants. In general there is no correlation between the pressure with which the sap is exuded from cut stems and the volume of sap flow. In some species relatively large volumes of sap may be exuded under a relatively low pressure, in others exactly the opposite relation may exist. Some of the woody plants in which bleeding from cut surfaces occurs,commonly ascribed in part or entirely to root pressure, are the sugar maple,box elder, dogwood, hornbeam (Ostrya), sycamore, birches, currant, and grape. Such exudation of sap under pressure usually occurs only during the seasons of the year when the plant bears no foliage, and most commonly in the early spring months.

The popular concept of a rise of sap in woody plants in the spring is largely based on the observation of such phenomena. Exudation of sap under pressure can also be demonstrated in many species of herbaceous plants especially if potted plants are used and the soil is flooded with water before the demonstration is attempted. The sap which moves through woody plants in the spring and, less abundantly, in the autumn under the influence of root pressures contains appreciable concentrations of carbohydrates as well as traces of mineral salts. In contrast with this is the condition of the xylem sap in the summer at which season it contains traces of mineral salts, but practically no soluble carbohydrates.It has sometimes been claimed that root pressures are adequate to account for the rise of water in herbaceous plants and in low shrubs or trees,and that they may even play a very considerable part in the ascent of sap in taller trees. While it is undoubtedly true that root pressure does, in some species of plants, under certain conditions, account for the movement of some water in an upward direction through plants, this process is inadequate to account for the translocation of any considerable amount of water through plants. In the first place there are many species in which this phenomenon has never been observed.

In the second place, the magnitude of the pressure developed is seldom sufficient to force water to the top of any except relatively low-growing species of plants. Neither is the rate of flow, as it occurs inmost species, anywhere near rapid enough to compensate for the known rates of transpiration. Finally, and this is perhaps the most fundamental objection to the idea that root pressures play a very prominent part in the ascent of sap in plants; they are usually negligible, in temperate regions at least,during the summer period when transpiration is most rapid. During periods of rapid transpiration, the cut surfaces of plants not only fail to exude sap, but will usually absorb water if it is supplied at the cut surface.

The trunks of many species of deciduous trees exhibit a marked seasonal variation in water content. A minimum water content is attained during middle or late summer and the trunks gradually fill up during the fall and spring. Often there is a temporary shrinkage in their water content during the winter months but the main period of diminishing trunk water content occurs in the spring and summer. Some authorities even speak of a seasonal rise and fall in the water table of a tree trunk, and consider that intercellular spaces as well as cells become occupied with water during this process. Root pressure undoubtedly accounts at least in part for such seasonal replenishments of the store of water in tree trunks.

The living cells of the xylem probably play an active part in this process. It is possible that there may be slowly ascending streams of water in the living cells of the xylem which operate entirely independently of the water columns in the vessels or tracheids. Hence the same portions of the xylem which at times contain gases may at other seasons be occupied by water. The trunks of many conifers show no marked seasonal variations in water content; this is possibly correlated with the lack of appreciable root pressures in such species.

The Cohesion of Water Theory. The leading advocate of this theory has been Dixon who has performed much of the experimental work upon which it is based.

A number of other workers, notably Askenasy and Renner have also contributed important theoretical and experimental evidence in support of this theory. Molecules of water, although ceaselessly in motion, are also attracted to each other by strong cohesive forces. In masses of liquid water the existence of such cohesive forces is not obvious, but when water is confined in long tubes of small diameter the existence of attractive forces between water molecules can often be demonstrated. If the water at the top of such a tube be subjected to a pull the resulting stress will be transmitted all along the column of water, due to the mutual attraction between the molecules.

The water conducting system of plants constitutes just such a system, enclosing continuous thread-like columns of water which extend from the top to the bottom of the plant. A stress applied at any point to this system will be propagated to all its parts. The application of such a stress stretches the water into taut threads, throwing it into a condition of tension, which is the equivalent of negative pressure. According to this theory forces develop in the upper parts of plants, and especially in the leaves, which cause the rise of water through the plant.

As evapouration proceeds from the walls of the mesophyll cells into the intercellular spaces the diffusion pressure deficit of the mesophyll cell walls increases. Water therefore moves into the walls from the adjacent protoplasm,this resulting in turn in the movement of water from the cell sap into the protoplasmic layer. The resulting increase in the diffusion pressure deficitof the cell sap is in turn propagated to the protoplasm and the cell sap in all parts of the cell. The osmotic pressures of the leaf cells are generally high enough that diffusion pressure deficits of sufficient magnitude to account for movement of water to the top of the plant can develop in them without a complete loss of turgor by the leaf cells.

If the cell is in direct contact with one of the branches of the xylem ducts which ramify throughout the lamina of the leaf, water will move from the vessel or tracheid into the cell this results in the development of tension in the water column terminating in this xylem element. The diffusion pressure deficit of the water in the conducting elements will be increased by the amount of this tension.

Water under a tension (negative pressure) of 10 atmos., for example, has a diffusion pressure just 10 atmos. less than that of pure water at the same temperature which is not under tension. In other words its diffusion pressure deficit is 10 atmos. If the cell from which evapouration is occurring is not in direct contact with a xylem element a gradient of diffusion pressure deficits, gradually increasing in magnitude from cell to cell in the direction in which the water is moving, is established between the xylem element and the cell from which evapouration is occurring.The tension developed in the conducting elements is transmitted along their entire length to their lower termination just back of the root tips and probably, very often at least, across the root tissues as well. The magnitude of the tension which develops in the xylem conduits is increased by conditions which favour a rate of water loss from the leaves considerably in excess of the rate at which water enters the roots from the soil. The movement of water across the cells of the root from the soil into the lower ends of the water-conducting elements must be considered as an integral part of the translocational process. At their lower terminations thexylem vessels or tracheids are in contact with the pericycle, or more rarely,directly with the endodermal cells of the root. Water moves from the adjacent root cells into the conducting elements because the tension (diffusion pressure deficit) developed in these elements exceeds the diffusion pressure deficit of the cells of the root.

When the tension in the water columns is relatively low a gradient of diffusion pressure deficits will be established across the root cells similar to those which are established in the leaf mesophyll cells when water is moving through them. The diffusion pressure deficits will increase from cell to cell along this gradient in the direction in which water is moving, *i.e.* from the periphery of the root towards the xylem.

Since, however, the osmotic pressures of root cells are usually lower than those of the mesophyll cells it is probable that the tension developed in the water columns often exceeds the highest osmotic pressures of any of the root cells through which water passes on its way from the soil to the xylem. Under such conditions the volume of water in the root cells may continue to diminish even after their turgor has been reduced to zero. As a result the walls of the cell are pulled inwards due to the adhesion between them and the contracting mass of water.

The counter pull exerted by the elastic walls of the cell will stretch the encompassed mass of water and throw it into a state of tension. Whenever a tension of sufficient magnitude has been generated in the conducting elements, therefore,the water in the root cells may also pass into a state of tension. Even under such conditions a gradient of diffusion pressure deficits would be established across the root, but the mechanism responsible for the development of this gradient would be more complicated than were no tension generated in the cells.

When the water in a cell is under tension its diffusion pressure deficit is equal to its osmotic pressure plus the tension to which the water is subjected. Regardless of the exact mechanism involved, the essential fact is that the development of tensions in the water columns induces the establishment of a gradient of diffusion pressure deficits, increasing consistently from cell to cell across the root from its peripheral cell layer to the conducting elements and water will move along this gradient from the epidermal layer to thexylem tissue.

Water enters the peripheral cells of a root whenever the diffusion pressure deficit of water in the soil is less than that of the water in the epidermal cells and root hairs in the absorbing zone of the root. In most trees practically all upward movement of water occurs in the vessels or tracheids of a few of the outermost annual rings of the sapwood and in some species, especially those with ring-porous wood is almost entirely confined to the outermost ring. As is well known, in some species of trees (beech, sycamore, etc.) the heartwood of the trunk may disappear completely by decay. The fact that such hollow trees continue to live and thrive is conclusive evidence that the heartwood plays no essential role in the upward movement of water in such species.

The water-conductive system of plants should be regarded as a unit system; the individual threads or columns of water not operating separately but each as a part of a more or less complicated meshwork. The air-plugged tracheids or vessel sections become merely islands in the meshwork of intact water columns, and the capacity of the tissues as a whole to conduct water, although somewhat impaired by the presence of air in some of the conducting elements, is by no means destroyed.

MAGNITUDE OF THE COHESIVE FORCE OF WATER

If water is to move through the xylem ducts as postulated by this theory, its cohesive force must be of sufficient magnitude to resist the stresses which are imposed upon it.The maximum height to which water ever moves in plants does not exceed 400 feet. This height is equivalent to about 2 atmos. which represents the maximum stress which the water columns could develop as a result of their own weight. Water, however, encounters a certain amount of resistance in moving through the conducting tissues. The greater the velocity of the current of water, the greater the resistance which it will encounter.

Dixon has shown that at the velocity corresponding to usual rates of transpiration this resistance is approximately equal to the pressure required to support water for the height of the plant. He used the wood of the yew for his determinations, in which higher values for resistance would be expected than in non-conifers in which most water traverses vessels rather than tracheids. For high rates of conduction Huber estimates the resistance to be several times as great as this. Accepting Dixons estimates as an average value,the required cohesive force of water must be doubled, giving a value of 24 atmos.

To this must be added the resistance normally encoun-tered by water in crossing the tissues of the root and the mesophyll cells of the leaves which is equivalent to only a few atmospheres. The estimated minimum cohesive force required to lift water to the tops of the tallest tree is therefore about 30 atmos. Even if we assume a resistance three times as great as that found by Dixon, the value of the necessary cohesive force becomes only about 50 atmos. Values experimentally determined for the cohesive force of water range up to 350 atmos. Dixon obtained values as high as 207 atmos. for sap centrifuged from branches of holly (hexaquifoliurn) The cohesive force of water is therefore much in excess of the minimum required for the lifting of water to the tops of even the tallest trees.

Although the estimates given in the preceding paragraph represent the minimum cohesive force actually required for the movement of water to the top of tall trees, many botanists believe that when an internal water deficit exists in plants, tensions as great as 100 atmos. or even more may develop in the water columns of at least some species of plants. One of the most ingenious methods of studying the cohesion of water employs the sporangia of the ferns for this purpose.

Around the edge of a fern sporangium occurs a ring of dead, thick walled cells, known as the annulus. As the sporangium matures water is gradually lost from the cells of the annulus by evapouration. Because of the powerful adhesion between the cell walls and water, the resulting shrinkage in the volume of water in each of these cells results in the thin outer wall of each cell being drawn inward, while the ends of the thicker lateral walls are pulled towards each other.

This results in tearing open the weak side ofthe sporangium, and eventually, in a complete inversion in the position of the annulus, exposing the spores on its outer surface. The annulus is now in a condition similar to that of a highly strained spring.

THE MOVEMENT OF WATER THROUGH THE PLANT

Eventually the cohesive force of the water in the cells of the annulus is exceeded, the water ruptures, and the annulus recoils suddenly into approximately its original position, scattering the spores into the atmosphere by its violent rebound. The cohesive force of the water in cells such as these can be measured by allowing the annuli to come to equilibrium with known vapour pressures, and determining the highest vapour pressure at which rupturing of the water occurs. The osmotic pressure corresponding to this vapour pressure is equal to the cohesive force of the water in the annulus cells. Such measurements indicate the cohesive force of the water in these cells to be, at a minimum, between 300 and 350 atmospheres.

The first experiment of this type was performed by Askenasy. As evapouration proceeds from the porous clay cup of this apparatus water moves up the vertical glass tube, followed by mercury from the reservoir. If the demonstration is successful the mercury will continue to rise above the level to which it will stand in a barometer. The water is now being pulled up the tube, a phenomenon which is possible only because of its cohesive force. The water in the tube is thus thrown into a state of tension which is transmitted to the mercury column below it, due to the relatively enormous adhesive force between water and mercury.The forces exerting this pull originate at the evapourating surface of the cup, and are due to the adhesive and cohesive forces operative in the maintenance of innumerable microscopic menisci in the pores of the clay cup. That succeeded in demonstrating a rise of mercury to a height of 226.6 cm. in such an apparatus. This is approximately three times as high as a column of mercury will be supported by atmospheric pressure acting alone. While the maximum tension developed in the water column in these experiments does not exceed two atmospheres, the demonstration graphically illustrates a physical system which is analogous to that which is believed to operate in the plant. Experiments similar to that just described have also been performed using small branches or twigs of plants as an evapourating surface instead of a porous clay surface.

These are attached at the upper end of a glass tube in place of the clay cup; otherwise the set-up for such an experiment is similar to that just described. While it has not been possible to obtain as great a rise of mercury in these experiments as when a porous clay evapourating surface isused, upward traction of mercury for distances well in excess of atmospheric pressure can be demonstrated if a suitable technique is followed.

DEVELOPMENT OF TENSION IN THE WATER COLUMNS

Convincing evidence that the water in the xylem vessels is often in a state of tension has been obtained by direct observations of vessels under a microscope. The stems of some species of herbaceous plants, especially the Cucurbits, are especially suitable for such observations.

Such a stem of an intact, rapidly transpiring plant can be fastened in position across the stage of a microscope and, by careful dissection, vessels examined individually. If one of the vessels under observation be jabbed with the point of a fine needle, an immediate jerking apart of the water column at the point of the rupture will be seen to occur, indicating that the water in the intact vessel was actually in a state of tension. Interesting evidence that the water in the xylem ducts of woody stemsis, at times at least, under tension has been obtained by means of an instrument known as the dendrograph. This is a self-recording instrument which measures variation in the diameters of tree trunks.

It is so constructed that its sensitivity is very great and that its recordings are not influenced by temperature effects upon the instrument. Dendographs are used principally to measure periodic variations in the diameter growth of trees. However, even in trees in which diameter growth has ceased, slight diurnal periodic variations in the diameter of trees have been found to be of regular occurrence. A record of the periodic variations in the diameter of a tree trunk for aperiod of several days at a season when little diameter growth was occurring. The trunk attained its minimum diameter during the transported through plants, be they the tallest of trees or herbs only a few feet in height.

It is also generally agreed that root pressures play a minorrole in this process at least under certain conditions and in some species.There is no doubt that living cells are involved in the phenomena generally described under the term root pressure and there is at least a possibility that living cells of the xylem are in some way essential to the maintenance or operation of the cohesion mechanism.

RATES OF MOVEMENT OF WATER THROUGH PLANTS

The rate which water ascends through the xylem ducts may vary from a movement so slow as to bealmost imperceptible to a speed of at least 75 cm. per minute. By rate is meant the length of the water column which will move past a given point in one minute. In general daily variations in the rate of ascent of water closely parallel daily variations in transpiration rate.

LATERAL MOVEMENT OF WATER

A cell to cell lateral movement of water in a radial direction undoubtedly occurs along the vascular rays in the stems of most species of plants. In woody stems there is probably also a lateral movement of water around the stem in a tangential direction. Except in trees in which the grain of the wood is twisted, the conducting vessels on one side of the tree generally connect with branches on that side of the treeat their upper extremity, and with roots on the same side of the tree at their lower extremity.

If no lateral movement of water occurred in woody stems,it would be expected that removal of the roots from one side of a tree would result in a dearth of water, or perhaps even death of the leaves or brancheson that side of the tree. Experiments have been performed on apple, peach, oak, and other woody species in which the roots on one side of the plant were removed in an attempt to determine whether or not lateral movement of the water occurs.Although the water content and growth of the plants treated in this manner diminished there was no difference in the moisture content of the leaves on the two sides of the tree. Neither did the leaves on the side from which the roots had been removed show any greater tendency to wilt on clear warm days than those on the other side. These results indicate very strongly that lateral movement of water occurs in woody stems, and that the water conductive system of plants acts as a unit system.

DOWNWARD MOVEMENT OF WATER

There are a number of experiments on record which indicate that downward movement of water can occur in stems. For example, Dixon found that if the tip of the leaf of a potato plant be cut off under an eosin solution the liquid will descend through the xylem tissue of the leaf, petiole, and stem, and eventually reach the under ground organs. In general, however, such an effect is to be expected only when an internal water deficit exists in a plant.

The cohesion theory of the movement of water will account equally well for the conduction of water in either the upward or the downward direction through the plant. If conditions are ever such that the diffusion pressure deficit of the cells of roots or some other basal organ of a plant exceeds that of the apical parts of the same plant a reversal of the direction of movement of water in the xylem will occur. Although downward movement of water can be readily induced in plants under certain experimental conditions, such movements of water are probably of infrequent occurrence in intact plants.

THE MAGNITUDE OF TRANSPIRATION

The Magnitude of Transpiration whether computed on the basis of a unit area of leaf surface, an entire plant or an acre of forest or crop plants, shows a great variation depending both on the plant species and environmental conditions to which they are exposed. The quantity of water transpired may be computed and compared for hourly daily, seasonal, or yearly periods. Because of the great fluctuations in transpiration rates with environmental conditions the citation of isolated values of transpiration rates is usually of very little physiological significance.

It is impossible to calculate accurately the transpiration rate of large plants such as trees from the known rate of transpiration of some of the leaves and an estimate of the area of all of the leaves on the tree. In the first place it is difficult to arrive at any accurate estimate of the aggregate leaf area of a large tree. In the second, there is a great variability in the transpiration rate among the different leaves. The rate of transpiration of the individual leaves varies depending upon their position, exposure, age, and internal physiological conditions.For example, Huber found that low twigs of a redwood tree transpired six times as rapidly as similar twigs from the height of 12 metres. Neither can it be assumed, as mentioned in the discussion of the potometer method, that the transpiration rate of leaves of branches which may have been removed from a tree and placed with their cut ends in water will be the same as their rate of transpiration before the branches were detached. Hence attempts to determine the transpiration rate for an entire tree from values determined for a few of the leaves on that tree can only result inapproximations. The same comments in general apply, although perhaps not so emphatically, to calculations of the total transpiration of fields of cultivated crops and natural vegetation areas based upon transpiration as determined for a limited number of individual plants.

If transpiration is measured for one minute intervals rates much in excess of this may be observed, but such rates are maintained only for relatively short periods of time of necessity approximations, although, not of course, without their utility in the consideration of certain types of problems. The results of such calculations indicate that sufficient water may transpire from maize plants during the course of a season to cover the field in which they are growing to a depth of 15 inches. Similarly Minckler has calculated that in a growing season water equivalent to 4.8 inches of rainfall may transpire from an acre of American elm trees and that an acre of red maples, growing in a very moist habitat, may lose water equivalent to 28.3 inches per acre. In evaluating all such data it must be remembered, however, that the magnitude of transpiration varies tremendously with the available soil water supply.

SIGNIFICANCE OF TRANSPIRATION

Opinions regarding the significance of transpiration have ranged all the way from those which would put it practically on a par with such processes as photosynthesis and respiration, to those which would relegate it to the category of a necessary evil. The principal roles which have been ascribed to the transpiration of plants can be summarized under the following three headings: Supposed Role in the Movement of Water.

It is often claimed or assumed that the movement of water through the plant requires the occurrence of transpiration from the leaves. That this concept is entirely erroneous will be clear from the discussion of the mechanism of the conduction of water through plants. Under conditions of high transpiration the movement of water through plants is, in general, more rapid than under conditions of low transpiration. The mechanism causing ascent of water through a plant operates in such a way that any decrease in the turgidity of the mesophyll cells favours a more rapid movement of water towards those cells. Hence a rapid transpiration rate, which invariably results in a considerable loss of turgidity by the mesophyll cells, usually speeds up the rate at which water ascends through the plant. However translocation of water to the extent that it is used in photosynthesis, growth or other metabolic processes continues even if the transpiration rate is negligible.

Supposed Role in the Absorption and Translocation of Mineral Salts. It has often been assumed that the more rapid the rate of transpiration, the greater the rate of absorption of mineral salts. This view implies that the dissolved mineral salts are swept into the plant along with the absorbed water, a postulation which ignores much evidence that the mechanisms operating in the absorption of water and the absorption of mineral salts are very different. The results of certain experiments do indicate that a somewhat larger quantity of mineral salts accumulates in plants under conditions favouring high transpiration than in similar plants growing under conditions of low transpiration although there is no actual proportionality between the volume of water absorbed and the quantity of mineral salts which pass into the plant.

Since even when plants grow under conditions favouring low transpiration rates they usually obtain an adequate quota of the various mineral salts, provided they are present in sufficient abundance in the soil, it is difficult to see that this effect is of any very great advantage to the plant. Realization is now fairly general that there is, at the most, only a slight correlation between the rate of transpiration and the rate of absorption of mineral salts from the soil.

It is also generally considered that transpiration plays an important part in the translocation of mineral salts from the root system to the top of the plant. That some upward translocation of mineral salts occurs in the transpiration stream is indisputable, but whether this is an important part of the total quantity transported is open to debate. There is considerable evidence that at least a part of the mineral salts which move through the plant are translocated in the phloem. The rate of transpiration can have little or no direct effect upon the movement of such salts. While high transpiration rates may mean a somewhat higher rate of transport of mineral salts through the plant than low ones it is doubtful if this is of great significance from the standpoint of the plant. Supposed Role in the Dissipation of Radiant Energy.

Leaves exposed to direct sunlight absorb large quantities of radiant energy which, unless dissipated in some other way, will be converted into heat energy and raise the temperature of the leaf. The possibilities of such an effect are indicated by the following approximate calculations: In direct noon-day summer sunlight the rate of receipt of solar energy is often 1.3 g.-cal. per square centimeter of leaf surface per minute and not uncommonly even greater. The proportion of this actually absorbed varies with the kind of leaf, but will be assumed to be only 50 per cent, which is a conservative estimate.

The incident radiant energy which is not absorbed by the leaf is all transmitted or reflected. Such a small proportion of the absorbed energy is used in photosynthesis that it can be neglected in a rough calculation. Hence about 0.65 g.-cal. of energy is absorbed per square centimeter of leaf surface per minute. If the mass of a square centimeter of a leaf is taken as 0.020 g. and its specific heat as 0.879g.–cal.2 the rise in temperature per minute would be 0.60.020 × 0.879 or about 370 °C. Since the thermal death point of most plant protoplasm lies. These are the actual values as determined for a sunflower leaf by Brown and Escombe.

It should not require more than two minutes to heat the leaves of most plants to a lethal temperature. Actual measurements of leaf temperatures, however, show that they seldom even approach their thermal death points. Leaf temperatures usually do not exceed atmospheric temperatures by more than a few degrees centigrade. Evidently some efficient energy-dissipating mechanism is at work which prevents the accumulation of heat energy in leaves. Since transpiration is an energy-consuming process it has often been assumed that it is in the evapouration of water from the leaves that most of the energy absorbed by them is dissipated. We might well Enquire, therefore regarding the possible efficiency of transpiration as an energy-dissipating process. The evapouration of a gram of water at 20 °C. requires 584 g.-cal. For the dissipation of 0.6 g.-cal. of heat, therefore, the evapouration of 0.0011 g. (0.65) of water per square centimeter of leaf surface per minute is required.

This is equivalent to 6.6 g. (0.0011 × 100 × 60) of water per square decimeter of leaf surface per hour, a rate of transpiration which is seldom if ever attained by plants for any sustained period of time. Evidently transpiration even when occurring at its maximum rate, can at the most account for a dissipation of only part of the radiant energy which is absorbed by leaves indirect sunlight.

The fact that transpiration is often inadequate in direct sunlight to account for the dissipation of all the absorbed radiant energy leads naturally to the question of whether it is at all essential for this process. Observations have shown that leaves in which the transpiration rate is greatly reduced, as for example those in which the stomates are plugged with vaseline, leaves in a wilted condition, or the leaves of plants in xeric habitats during dry seasons in which the occurrence of any appreciable amount of transpiration is precluded by the lack of soil water, seldom have temperatures which are any where near the thermal death point, even when exposed to direct sunlight.

Although transpiration often accounts for the dissipation of some or even most of the absorbed radiation, in so doing it apparently plays no essential role since absorbed radiation can be dissipated by purely physical means. As soon as the temperature of a leaf exceeds that of the surrounding atmosphere, it will lose heat to the atmosphere in the same way that any other object heated above its environmental temperature does, *i.e.* by one or more of the purely physical processes of conduction, radiation, and convection.

The term thermalemissivity is frequently applied to this physical loss of heat energy by leaves and other objects. Actually, instead of benefiting plants, transpiration may often be detrimental. Under conditions of deficient soil water or of high transpiration rates even when the soil water supply is adequate, this process results in adiminution in the water content of a plant and in the turgidity of its cells. Prolonged drought conditions ultimately result in a severe desiccation withthe consequent death of all except the most drought resistant species.

When the diminution in water content is less severe, a train of other effects such as a decrease in the turgidity of the cells, stomatal closure and reduction in rate or cessation of photosynthesis are induced, all of which have the end result of checking the growth of the plant. It is probably true that lack of water is more often the limiting factor in plant growth than any other single factor. Furthermore deficiency of water is probably responsible for the death of more plants under natural or even cultural conditions than any other single cause.

The fundamental effects of transpiration upon the plant are not to besought in any hypothetical advantages of the process to the plant, but in its very real and readily ascertainable influences upon the water relations of plant cells and tissues, and through these its effects upon other plant processes. In spite of the fact that transpiration may be regarded in a sense as an incidental phenomenon, its indirect influence upon the metabolic processes of plants is a profound one. It is this fact, more than any other, which justifies intensive and critical studies of this process.

THE LOSS OF WATER FROM PLANTS IN LIQUID FORM

If a pot of young oat plants be copiously watered and then enclosed in a bell jar, in a relatively short time a slow exudation of water begins at the tip of each leaf. The drops which form at the leaf apices gradually enlarge and eventually may run down the side of the leaf or fall off. This process of the escape of liquid water from uninjured plants is called guttation. It is of very general occurrence in plants, baying been recorded in plants of more than 300 genera although there are many species in which it has not been observed. Guttation occurs most frequently and abundantly under conditions which favour rapid absorption of water by the roots but which result in a reduced rate of transpiration. In most temperate regions such conditions occur most frequently during the late spring when there is often an alternation between relatively cool nights and relatively warm days. Guttation is frequently observed at that season, usually taking place at night or during the early hours of the morning.

The drops of guttation water which form at the tips of grass blades and the tips or edges of the leaves of other plants are often erroneously considered to be dew. The process of guttation occurs from specialized structures known as hydathodes. These structures are also sometimes referred to as water stomates or water pores. A typical hydathode consists of an enlarged stomate-like opening below which is a rather large. Guttation from tomato leaves. Photograph, courtesy of Dr. J. H. Gourley chamber, bordered by a mass of thin-walled, loosely arranged cells called the epithem. The xylem elements of a vascular bundle terminate just below the epithem.

The exudation of water through hydathodes is considered to result from develops in the 'ap of the xylem elements, and not to any locally developed pressure in the hydathode itself. This pressure is generally believed to be identical with the so-called root pressure. It is supposed that the water is forced from the vessels through the intercellular spaces of the epithem layer and out of the plant through the pore of the hydathode. This water is not pure but contains traces of sugars and other solutes. While the volume of water exuded by most temperate region plants in guttation is usually insignificant, some tropical species lose large quantities in this process.

A young leaf of Colacasia nymphaefolia, a native of India, has been observed to lose as much as 100 cc. of liquid water in a single night by guttation. Glands are found on leaves, flower parts and other organs of the plant. Many of them represent modified epidermal hairs or other epidermal cells. Certain types of glands secrete water or, more accurately, a dilute sap. These have sometimes been classed as hydathodes, but it seems better to restrict this term to the type of structure previously described under this name epithem. The exudation of water by glands is generally called secretion and is apparently caused by forces which develop within the gland itself, and not by a pressure developed in the sap of the water conducting tissue as appears to be true of hydathodes. Glands which secrete water or dilute saps are not closely connected with xylem elements as are the hydathodes which secrete other substances as well as water. Among these are sugars (as in nectar), calcium salts, resins, volatile oils, and enzymes.

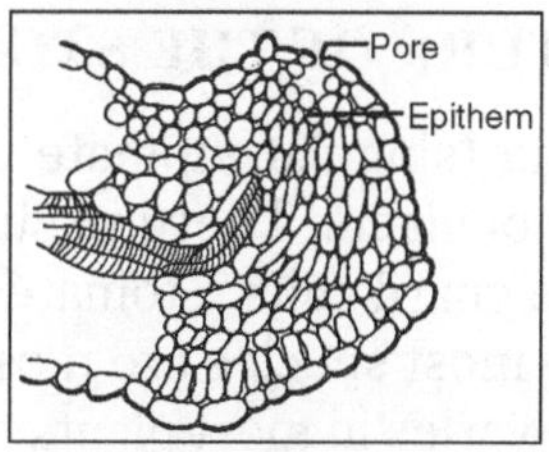

Fig. Hydathode at the Margin of a Tomato Leaf as Seen in Sectional view (Semi-diagrammatic).

Note: Termination of Vessel Just back of the Epiterm.

If the stem of an herbaceous plant be cut or broken a slow exudation of sap often occurs from the ruptured stump. A similar phenomenon can often be observed in the cut or broken stems or branches of woody plants, especially in the spring of the year. The exudation of sap from holes bored into maple trees is a good example. This phenomenon has long been called bleeding. Under certain conditions large quantities of dilute sap may be lost in this process.

A single vigorous grape vine may lose as much as one litre of sapin a day, while under favourable conditions a sugar maple tree will yield from five to six litres in twenty-four hours. In some species bleeding appears to be due to a root pressure developed in the xylem elements, in others to pressures developed in the phloem, and instill others to locally developed pressures in other tissues. The significance of the various internal pressures which develop in the conductive tissues of plants will be discussed later, particularly in relation to the problems of the translocation of water and solutes in plants.

THE STOMATAL MECHANISM

The most important physiological fact about the stomates is that they are sometimes open and sometimes closed. When open they serve as the principal pathways through which gaseous exchanges take place between the intercellular spaces of the leaf and the surrounding atmosphere. When closed all gaseous exchanges between a leaf and its environment are greatly retarded.The gases of greatest physiological importance which enter or depart from a leaf principally through the stomates are oxygen, carbon dioxide and water vapour. The movement of gases through the stomates in either direction is primarily a diffusion phenomenon, although under certain conditions, as for example when the intercellular spaces are alternately compressed and expanded by bending of the leaf in a high wind, alternate outward and inward mass movements of gases may occur.

Although the stomates are the principal portals through which entry and escape of gases takes place, the fact shouldn't be overlooked that at least small amounts of these three gases pass directly through the epidermis and cuticular layers of all leaves. In submerged aquatics all gaseous exchanges between the plant and its environment occur through the epidermis.

STRUCTURE OF THE STOMATES

The stomates or stomata (singular stomate or stoma) are minute pores which occur in the epidermis of plants. They are surrounded by two specialized epidermal cells known as the guard cells. Stomates may occur on any part of a plant except the roots, but in most species are most abundant upon the leaves. The size of the stomatal pore varies in most plants depending upon the turgidity of the guard cells and often, especially at night, it is entirely closed. The structure of stomatal apparatus shows marked variations in detail indifferent species of plants, but the essential feature of a pre between two modified epidermal cells is common to all species of vascular plants. Guard cells which are roughly kidney or bean-shaped as seen in surface view are typical of more species of plants than any other kind. In some species the epidermal cells bordering on the guard cells are different in con figuration from other cells in the same tissue; these are called subsidiary cells or accessory cells. In many species of the grass and sedge families the guard cells are distinctly elongate.

In most species of conifers, and in certain other species, stomates are of the sunken type The guard cells differ from the other epidermal cells in that they contain plastids which are green in colour. These have often been assumed to bechloroplasts, and are frequently so-called, but the green pigment present does not seem to be true chlorophyll as photosynthesis apparently does not occurring the guard cells.

SIZE AND DISTRIBUTION OF STOMATES

The size of the stomatal pore varies greatly according to the species of plant, and somewhat even among the individual stomates on any one plant. They are always very minute however, their dimensions being expressed in terms of microns. Minute as these openings appear to be from a human scale of values they are enormous when compared with the size of the gas molecules which diffuse through them. The calculated diameter of a water molecule is 0.000454. More than 2,000 water molecules would have to be placed side by side to measure a distance.

The molecules of both carbon dioxide and oxygen are larger than water molecules. Since the stomatal diameters usually are considerably in excess, it is evident that the stomates afford relatively enormous portals to the gas molecules which diffuse through them. In general the number of stomates present in the epidermis of leaves may range from a few thousand to over a hundred thousand per square centimeter the exact number depending upon the species and upon the environmental conditions under which the leaf has developed.

A single maize plant has been estimated to bear from 140 to 240 million stomates, while the number on a large tree could be expressed only by a figure of astronomical dimensions. However, marked deviations from such average values are possible for any species, depending upon the environmental conditions under whichthe leaves have developed. The number of stomates per unit area of leaf surface may be quite different on leaves of two plants of the same species if one grew in a greenhouse, and the other grew in the open, or upon the leaves of plants of the same species which have developed during different seasons. Stomates occur in both the upper and lower epidermis of many species of plants. In numerous others, especially woody species, they are confined to the lower epidermis. Even in those species in which stomates occur on both surfaces of the leaf they are commonly, but not always, more abundant in the lower epidermis. In floating leaves such as those of the water-lily, stomates occur only in the upper epidermis.

Species in which the stomates are relatively small usually have more per unit area than species in which the stomates are relatively large. The number of stomates per unit area usually varies on different leaves of the same plant, and even in different parts of the same leaf. On individual leaves it appears to be generally true that the greatest number of stomates per unit area is at the tip, the lowest towards the base, while the middle portion of a leaf shows a frequency midway between these two extremes.

As a rule, the higher the point of attachment of a leaf on the stem of a plant the greater the number of stomates per unit area. In general no correlation has been found between transpiration rates and either the size or distribution of stomates, other factors being much more important in determining the rate of loss of water-vapour from the intercellular spaces.

PRINCIPLES GOVERNING DIFFUSION OF GASES THROUGH THE STOMATES

Since virtually all gaseous exchanges between the intercellular spaces and the atmosphere take place through the stomates, the problem of the diffusive capacity of the stomatal pores is an important one. Although the aggregatearea of the fully open stomates on a single leaf is seldom more than 3 per cent of the leaf area and is often as low as i per cent, the rate of water loss from the leaves (*i.e.* loss per unit area in a unit time) in many species maybe, under favourable conditions, as much as 50 per cent of the evapouration from an exposed water surface of comparable dimensions.

Stlfelt (1932) records that transpiration from a leaf of birch (Betula pubescens) under the most favourable conditions may be 65 per cent of the evapouration from a comparable evapourating surface, but this is probably an unusually high value. It is therefore evident that water-vapour often diffuses through the stomates at rates ranging up to at least 50 times greater than it diffuses away from an equal area of exposed evapourating surface.

This unexpectedly high diffusive capacity of the stomates is intelligible in terms of the results of studies upon the principles of diffusion through small apertures. The classical investigation of this problem was made by Brown and Escombe who studied the rate of diffusion of carbon dioxide through tubes of known dimensions. They first made the important discovery that if a septum which had been pierced with a small circular aperture was interposed across a column of diffusing gas, that the rate at which carbon dioxide diffused through this aperture was much greater than the rate at which it passed through an equal area of the unobstructed tube. Since the problem at present under consideration is the diffusion of water-vapour rather than carbon dioxide through small apertures will be used to illustrate more fully the principles regarding the diffusion of gases through small openings. These data were obtained by sealing thin septa, through the centre of which were cut round openings of known dimensions, across the circular mouths of small bottles which had previously been filled to a certain level with water, and then determining the loss in weight of each bottle after all of them had been kept under uniform conditions for the same period of time.

In other experiments, Sayre found that diffusion of water-vapour through small openings of elliptical shape is also more nearly proportional to their perimeters than to their areas.Hence the perimeter law for the diffusion of gases also holds for openings of other than circular shape. An important inference which can be drawn from this finding is that a stomate attains almost its maximum diffusive capacity long before it is fully open, since the perimeter of a stomatal pore does not increase greatly after the aperture between the guard cells has widened perceptibly.

The fact that the diffusion of gases through small apertures is much more nearly proportional to their perimeters than to their areas is due to the more rapid diffusion of the molecules through an opening at the peripherythan through its centre. A molecule of the diffusing gas at the rim of the hole is surrounded on all sides by other molecules moving in all directions.

The concentration of diffusing molecules is much less in an outward direction from the rim, however, than towards the centre of the opening. Furthermore, the diffusion gradient is much steeper from the rim outwards than at points above the opening. Hence the number of molecules escaping through the hole from a given point near the rim per unit time interval will greatly exceed the number escaping from a point near the centre of the opening. Reduction in the area of any circular or elliptical opening results in increasing its perimeter relative to its area.

Hence in small apertures such alarge proportion of the diffusion occurs at the edge of the opening that the effect of any diffusion which occurs through its centre is almost if not completely obscured, and measurements show the rate of diffusion through the small apertures to be essentially proportional to their perimeters. In fact in an opening 0.5 cm. in diameter, for example, a large part of the centre of the aperture can be entirely blocked off without any very great effect upon the rate of diffusion of a gas through it.

That the diffusion of gases through small apertures occurs largely at the rim of the hole can be demonstrated visually by means of a simple experiment. A cylindrical glass vessel is partly filled with a gelatin sol which is allowed to gel. A small opening is punched through the centre of a disk of celluloid which is then inserted into the position on top of the gelatin,the gap between the edge of the disk and the wall of the cylinder being sealed with melted vaseline in order to prevent leaks. A solution of potassiumpermanganate, which is purple in colour, is now introduced into the vessel ontop of the disk.

After a few hours the molecules of potassiumper-manganate, after diffusing through the aperture, become distributed in such a way as to occupy a hemispherical zone below the opening. The hemispherical distribution assumed by molecules which are diffusing through a small opening is often spoken of as a diffusion shell. Gases also form diffusion shells above small pores but their configuration is often relatively unstable due to their ready distortion by wind or convection currents. Actually when molecules are diffusing through a small orifice there is formed not one, but two diffusion shells, one on each side of the diaphragm which is pierced with the opening.

The experiment described above permits discernment of the hemispherical pattern of diffusion only on the receiving side of the system.There is also a second shell, a mirror image of the first adjacent to the aperture on the side of the system from which diffusion is occurring. The deep colour of the permanganate solution prevents visible detection of this second shell in the experiment described above.

Diffusion of gases from or into a leaf through the stomates involves a much more complex system than is represented by a septum pierced by a single aperture. Diffusion is occurring not through a single opening of potassium permanganate upon diffusion. Experiments into gelatin have been performed in analogous physical systems in which multiperforate instead of uniperforate diaphragms have been used.

The diffusion per pore increases with increase in the distance apart of the apertures, although not proportionately. With pores of this diameter (0.3 mm.) nearly the maximum diffusive capacity is attained when they are spaced at intervals of 20 diameters.

The closer together the pores in a septum the greater the overlapping of the diffusion shells. The molecules diffusing through each aperture invade in part the zones into whichthe molecules passing through neighbouring pores would diffuse were each opening the only one in the septum. Hence the diffusion gradients are less steep than they would be were there only one pore in the septum.As a result the rate of diffusion through each pore is less than it would be through a pore of equal dimensions in a uniperforate septum. The loss of water-vapour by diffusion per septum decreases with increase in the distance between the openings, *i.e.* with decrease in the number ofpores per septum. The decrease in diffusive capacity is not however, in proportion to the reduction in the aggregate area of the pores in the septum. For example, when the pores are spaced 5 diameters apart their aggregate area is only 3.38 per cent of the septum area, yet diffusion through them was 62 per cent of that through an open bottle with a mouth of the same area as the septum.

The dimensions of the pores (0.3 mm. diameter, 0.0707 mm^2 area) used in the experiments just discussed are much greater than those of the average stomate.What about the diffusive capacity of still smaller pores? Huber has shown that the smaller and more numerous the pores, aggregate pore area being constant., the greater the diffusion per septum. If equal-sized pores are spaced equidistantly, and represent the same proportionate area of the septum, they will be the same distance apart in terms of their diameters, regardless of pore size. Hubers results indicate that the smaller the pores for a given aggregate pore area the more nearly the diffusion through the septum will approach that of a free evapourating surface. In the sunflower and undoubtedly also in some other species the aggregate area of the stomatal pores is about 3 per cent of the leaf area.

If diffusion through pores 0.05 mm. in diameter,occupying 3.2 per cent of the area of the septum, is 72 per cent of that from a free evapourating surface, diffusion through the much smaller stomates of this species should be proportionately greater. In such species it seems probable that the aggregate diffusive capacity of the stomates mayclosely approach that of a free evapourating surface. Even in species in which the total area of the stomatal pores is less than in the sunflower their aggregate diffusive capacity is very large relative to their area. The relatively high rates of water loss which occur from leaves in proportion to the area of the stomatal pores are thus explicable in terms of the principles of diffusion through multiperforate septa.

The theoretical diffusive capacity of the stomates is more than adequate to account for known rates of transpiration. This is shown by the calculations of Brown and Escombe for sunflower leaves. Assuming a leaf temperature of 200 °C., the intercellular spaces to be saturated with water vapour at that temperature (vapour pressure 17.54 mm. Hg), and the vapour pressure of the atmosphere to be one-fourth of this value, their computations show that transpiration would occur at the rate of 17 g. of water-vapour per square decimeter of leaf surface per hour. This is several times greater thanthe maximum rate of transpiration ever recorded for a sunflower plant.

Evidently some other factor than the diffusive capacity of the stomates has a limiting effect upon transpiration whenever the stomates are widely open. The foregoing discussion of the diffusion of gases through small openings has been based on the assumption that diffusion is occurring into a quiet atmosphere. If an air current is blowing across the surface of a multiperforate septum with a sufficient velocity to prevent the formation of diffusion shells the result is a steeper diffusion gradient and hence a greater rate of diffusion of gas through the pores of the septum. Hence, when subjected to air movements, the diffusion of water through multipe-rforate septa or the stomates may be even greater than the preceding discussion has indicated. Within limits increase in the velocity of the wind results in a progressive increase in the rate of diffusion through a multiperforate septum.

THE MECHANISM OF THE OPENING AND CLOSING OF THE STOMATES

The degree of stomatal opening is influenced both by changes in the turgor of the guard cells and by changes in the turgor of the epidermal cells, although the former usually play a predominant role. In general an increase in the turgor of the guard cells relative to that of the epidermal cells leads to a widening of the stomatal aperture, and vice versa. The greater this turgor difference, the wider the stomatal aperture. The mechanism of the effect of changes in the turgor of the guard cells upon the size of the stomatal aperture varies with the structure, form, and position of the stomates. In one type of guard cell, found in many different species of plants, the cell wall is thicker on the side bordering the stomatal pore than on the side bordering the epidermal cells.

With an increase in turgor the thinner walls of the guard cells are stretched more than the thicker; this causes the thicker-walled sides to assume a concave shape and results in the appearance of a gap the stomatal pore between the two guard cells. Opening of the stomates typical of the grass and sedge families appears to be due to swelling of the ends of the guard cells thus separating the abutting walls of the middle portion of the two adjacent guard cells. In the sunken stomates typical of conifers opening of the stomates seems to result largely from a change in the shape of the guard cells due to an increased turgor which is unaccompanied by any appreciable stretching of the walls. These various types of stomatal mechanisms are discussed by Copeland.

The three principal factors which influence the opening and closing of the stomates are:

1. Light,
2. The internal water relations of the leaf, and
3. Temperature.

INFLUENCE OF THE LIGHT FACTOR IN STOMATAL OPENING AND CLOSING

The most familiar of all stomatal reactions is their response to light. Unless other conditions, to be discussed later, are limiting, the stomates of most species open when exposed to light, and close upon the failure of illumination. Most commonly, therefore, the stomates are open in the daytime and closed at night, although there are many exceptions to this statement. The sensitivity of stomates to the light factor probably varies considerably according to species.

Within limits the stomates appear to respond quantit-atively to the amount of light absorbed. Stomatal opening apparently will occur in all wavelengths of the visible spectrum, although the influence of radiations in the red region appears to be weaker than the influence of other wavelengths of the visible spectrum. Upon the cessation of illumination the stomates usually begin to close. Generally this is a gradual process and, according to Stiil felt the greater the quantity of light which has been absorbed during the course of a day, the longer it takes for the completion of stomatal closure.

The work of Sayre Scarth and others indicates that the mechanism whereby light brings about stomatal opening and the mechanism whereby its absence causes stomatal closure are primarily osmotic, but are conditioned by changes in the H-ion concentration of the guard cells. Illumination of the guard cells has been found to result in a decrease in their H-ion concentration; failure of illumination in an increase. Thus Scarth found the pH of the guard cells of the Wandering Jew (Zebrina pendula) to range from 5.0 or less in the dark to between 6.o and 7.4 in the light.

The other cells of the leaf do not change appreciably in H-ion concentration upon the advent or failure of illumination, suggesting that the guard cells are less effectively buffered than the other leaf cells. Although several explanations have been offered to account for the effects of light and darkness on the pH of the guard cells none of them are strongly supported by experimental evidence and they will not be discussed. The guard cells apparently always contain starch, but the quantity present is not constant from one hour of the day to the next. Sayre has shown that the starch content of the guard cells is at its maximum during the night, decreases rapidly during the daylight hours, and increases again towards evening.

The sugar content of the guard cells bears a reciprocal relationship to the starch content; when the latter is high, the sugar content is low, and vice versa.

These reciprocal changes are apparently the result of reversible reactions in which the total amount of carbohydrate involved doesn't vary greatly. Conversion of starch to sugar and of sugar to starch results from the action of the complex of enzymes known as diastase. Increase in the pH of the guard cells such as occurs upon the incidence of light, appears to favour the hydrolytic (starch to sugar) action of diastase. On the contrary a decrease in their pH, such as occurs in the evening, favours the synthetic action of this enzyme whereby sugar is converted into starch.

Increase in the sugar concentration of the guard cells results in an increase in their osmotic pressure while a decrease in their sugar concentration has the opposite effect. That such changes in the osmotic pressure of the guard cells actually occur has been shown by many investigations. The diurnal changes in the osmotic pressures at incipient plasmolysis of the guard cells and epidermal cells. In general the osmotic pressure of the guard cells is usually relatively high during the daylight hours, and relatively low at night.

The osmotic pressure of the epidermal cells does not change appreciably during the course of the day and approximates that of the guard cells at night. Increase in the osmotic pressure of the guard cells in the morning results in an increase in their diffusion pressure deficit relative to that of the contiguous cells. Water therefore moves into the guard cells, increasing their turgor, which in turn leads to a widening of the stomatal aperture. Similarly a decrease in the osmotic pressure of the guard cells results in a diminution of their turgor and a narrowing of the stomatal aperture. However, certain facts suggest that the effects of light upon stomates cannot be interpreted entirely in terms of the osmotic mechanism just described. One of these is the rapidity with which stomatal opening occurs. In some species at least the stomates open within less than a minute after exposure to light. It is difficult to visualize such a rapid action in terms of an enzymatic reaction, since usually such reactions occur at a relatively slow rate. Some investigators consider therefore that light exerts a direct effect upon the guard cells in addition to its indirect influence upon the magnitude of their osmotic pressure. Little if anything is known, however, concerning the mechanism of any such reaction.

9

Plant Anatomy and Plant Physiology

The origin of mans first conscious interest in plants long antedates recorded history. Agriculture had already become a highly developed art thousands of years before any experimental study of plant processes began. Consequently, there had grown up a vast body of traditional plant lore which passed orally from father to son, generation after generation. This practical knowledge of plants developed over many centuries as a result of countless, mostly involuntary trial and error experiences and innumerable observations of plant behaviour under all kinds of circumstances.

Which of this mass of mingled facts and beliefs regarding plants is alive in the consciousness of the common man today. The closer he lives to the soil and the more familiar he is with traditional plant lore, the more it influences his thoughts and actions. Many of these customarily accepted beliefs are essentially sound and most of them contain elements of truth. Others, however, are entirely erroneous, and not a few are tempered with superstitions, so me of which have an unbroken lineage back to the days of witch-doctors and savagery. No reputable botanist has held for generations, for example, that plants obtain their food from the soil, yet this and many other fallacious beliefs are still widely entertained among the general population.

The value of practical information about plants should not be underrated, since its perpetuation in the mind of man permitted the development of agriculture to a high plane as a practical art before any widespread investigation of plants from a scientific point of view was undertaken. Nevertheless, traditional plant lore is not only often inadequate, but is riddled with misconceptions, and often suffers from points of view which are inherently stultifying to the acquisition of further knowledge. The layman, for example, often personifies plants in an attempt to explain their behaviour. Man has desires and foresight and it is often assumed, either consciously or tacitly, that plants are similarly endowed. To many, for example, the statement that roots grow downwards in search of water, or that stems grow upwards in order to reach the light are accepted as adequate explanations of plant behaviour.Mans knowledge that water and light are essential to plants is not evidence that plants are similarly aware of these facts. To assume that plants realise their needs and are able to act in conformity with their requirements is equivalent to crediting them with a high order of intelligence. Explanations of plant behaviour are commonly encountered in which purposeful action on the part of plants is tacitly or deliberately implied, although there is no justification for the adoption of such a point of view. Furthermore, the layman seldom pursues his quest for information about plants beyond the stage of observation, while the scientist frequently does.

Observation has suggested, for example, that light is necessary for the continued existence of plants. To one who is scientifically minded, either by instinct or training, the obvious next step is to test this postulate experimentally. If the suggested hypothesis is substantiated by experiment it is tentatively accepted as a theory. Theories such as those proposed in explanation of the processes or reactions of plants, together with the experimental results which are considered to support them, are usually published in a scientific journal or monograph and thus exposed to evaluation and further experimental testing by other scientists.

Continued experimentation may lead to substantiation, rejection, or modification of the theory as originally proposed. Modification would undoubtedly be the fate of the hypothesis used as an example, since sooner or later some investigator would find that non-green plants can thrive in the absence of light. Experiments often raise more questions than they answer. New approaches to the problem under consideration as well as desirable new lines of inquiry are constantly opening up to the alert investigator. In this way experimentation leads to more experimentation, more facts accumulate, and more theories are proposed. Some of the suggested hypotheses are confirmed, others are rejected, and still others are modified. Most of them, sound or fallacious, in turn suggest further observation and experimentation.

As a result of such endless and painstaking labours there is slowly built up that vast, complex, and ever changing body of knowledge which we refer to as a science. The system of subjecting all hypothetical explanations of natural phenomena to experimentation is the essence of the scientific method. It is the characteristic methodology of all sciences. Progressive modification of accepted concepts in the light of new experimental findings continually increases the soundness of scientific generalizations. Thus there are incorporated into any science theories and generalizations in various stages of acceptance.

Some stand upon such a firm substructure of facts that they are accepted by all authorities in the field. Others, less securely supported by experimental results, are subscribed to by some but rejected by other workers. Finally, in any science there are always some theories which are so dubious that they find only a few advocates. Furthermore, some of the theories now widely held sooner or later will be discarded as a result of new findings or as a consequence of different interpretations of facts already known.However, not all scientists are in agreement regarding the interpretation of the same sets of facts. Although this state of affairs is entirely consistent with the spirit of scientific research, it is frequently puzzling to students and laymen. Differences of opinion regarding the hypotheses which suitably explain scientific phenomena are most likely to arise when experimental data are inadequate. Disagreements regarding the interpretation of experiments and observations are often inevitable steps in the clarification of scientific generalizations.

Controversies usually focus attention upon gaps in our factual information. Frequently, therefore, they are stimulating to research, and often lead to a further enrichment of human knowledge. Without the knowledge of plant processes which has been slowly accumulated through more than two centuries of observation, experimentation, and critical evaluation by numerous workers in all parts of the world, this book could never have been written. In spite of the patient labours of these many workers vast gaps still exist in our understanding of the physiology of plants gaps which are reflected in the necessarily inadequate treatment of many topics in this book. The future of this science and all others lies in the hands of the front-line investigators who wage a continual struggle for the extension of human knowledge.

POSSESSIONS OF SOLUTIONS

Water is the most abundant compound present in all physiologically active plant cells. The water which occurs in a liquid state in plant cells invariably contains other substances dissolved in it and usually also contains dispersed particles which are not in true solution. When the particles dispersed throughout the water are within a certain range of sizes the system of water plus particles falls into the category of colloidal systems. The complicated dynamics of living systems can be largely interpreted in terms of the physicochemical properties of solutions and colloidal systems, one component of which is water, although it should not be inferred that non-aqueous solutions and colloidal systems are entirely absent in living organisms.

Similarly liquid water never occurs in the pure state in the natural environment of living organisms. The water of streams, lakes, and oceans invariably contains various substances in solution and usually in the form of dispersed particles as well. This is likewise true of the soil water. Even raindrops, the products of natural distillation, contain gases which have dissolved in them from the atmosphere.

COMMON NATURE OF SOLUTIONS

Simple solutions are systems in which one component (the solute) is dispersed throughout the other (the solvent) in the form of molecules or molecules and ions. Theoretically the solvent may be a gas, a solid, or a liquid, but solutions in which the solvent is a liquid are by far the most important in living organisms. Except in extremely concentrated solutions the average distance between the solute particles is usually very great relative to their size.

Naturally occurring solutions, whether in living organisms or their environment, usually contain a number of different solutes and are often exceedingly complex. Water is the commonest and most important of all solvents both in the inorganic world and in the realm of living organisms. The further discussion will be principally in terms of aqueous solutions.

SOLUTIONS OF A GAS IN A LIQUID

The water present in living organisms usually contains dissolved gases. Those most commonly present are carbon dioxide, oxygen, and nitrogen. Practically all the water in the environment of organisms in oceans, lakes, streams, soils, and raindrops also contains these gases in solution, and sometimes others as well.

A given volume of water or any other liquid will hold only a limited quantity of gas in solution at a given temperature. When no more of a certain gas can be dissolved in a liquid it is said to be saturated with respect to that gas. Gases vary widely in their solubility in water, but in general fall in to two groups: those which are sparingly soluble and those which are highly soluble.

When only a small fraction of a unit volume of a gas will dissolve in a unit volume of water the gas is classified in the sparingly soluble group. When from one to many unit volumes of a gas will dissolve in one unit volume of water it is classified in the highly soluble group. Oxygen, hydrogen, and nitrogen are familiar examples of gases belonging to the first group, while carbon dioxide, ammonia, and hydrogen chloride are examples of the second. When gases are highly soluble in water it is usually evidence that a chemical reaction takes place between the gas and water.

There actions between water and carbon dioxide and water and ammonia are indicated in the following equations:

$$CO_2 + H_2O \rightarrow H_2CO_3NH_3 + H_2O + 1NH_4OH$$

In a solution of such gases not only are molecules of the gas present, but also molecules of a compound formed by the reaction of the gas with water. The apparently great solubility of gases such as these is due to the formation of relatively soluble compounds by the reaction of the gas with the water.

Increase in pressure of a gas above a liquid increases the solubility, *i.e.* the concentration of that gas in the liquid. For sparingly soluble gases the increase in solubility is directly proportional to the increase in pressure (Law of Henry). This principle also holds qualitatively, but not quantitatively, for the very soluble gases.

When a mixture of several gases is maintained over water, each dissolves independently and in accordance with the gaseous pressure (partial pressure) which it exerts against the surface of the water (Law of Dalton). This principle holds strictly only for those gases which are slightly soluble in water. For example about one-fifth of the atmospheric pressure is due to the oxygen present. Assuming an atmospheric pressure of 760 mm. Hg (the value at standard conditions), the pressure due to the oxygen is equivalent to about 152 mm. Hg. The quantity of oxygen which dissolves in water exposed to the air is the same as that which would dissolve if oxygen only at a pressure of 152 mm. Hg occupied the space over the water.

SOLUTIONS OF A LIQUID IN A LIQUID

In general solutions of a liquid in a liquid fall into two classes: those in which the liquids are freely miscible with each other in all proportions and those in which each liquid reaches a definite point of saturation in the other. Alcohol and water, for example, mix with each other in all proportions; such a solution is an example of the first mentioned type. Many oily liquids also are miscible with each other in all proportions. A familiar example is the solution of lubricating oil in gasoline.

In solutions of this kind the liquid present in excess is usually considered to be the solvent. In a 50 per cent solution of alcohol and water, either liquid could be considered the solvent, or either could be considered the solute. Ether, chloroform, and many other liquids are sparingly soluble in water. After water and ether, for example, are shaken together in a flask, two distinct layers of liquid separate upon standing. The upper layer consists of the lighter ether, saturated with water, while the lower consists of water saturated with ether. In both layers, however, the concentration of solute present at saturation is small. In the upper layer ether is the solvent, water the solute; the converse is true of the lower layer.

SOLUTIONS OF A SOLID IN A LIQUID

This is by far the most familiar type of solution and in many respects the most important. Substances in the solid state vary greatly in their solubility in water, ranging all the way from those which are virtually insoluble to those which are extremely soluble. There is usually a limit to the amount of any solute which can be dissolved in a given volume of water at a given temperature. When this limiting concentration is reached the solution is said to be saturated, following the same terminology used with solutions of gases and liquids in liquids. It is almost impossible to prepare a true saturated solution of any solute unless some of it is also present in the solid state.

Under certain conditions, and only when none of the undissolved solid is present, a super-saturated solution can be prepared; that is, a solution in which the concentration of the solute is greater than that in the saturated solution. If a fragment of the solid solute be added to such a solution the excess solute usually crystallizes out immediately, and the concentration of the solution decreases to that usually present at saturation. Usually increase in temperature increases the solubility of a solid in a liquid but there are some exceptions to this principle. The solubility of common salt (NaCl) in water, for example, is only slightly influenced by temperature, while the solubility of many calcium compounds is decreased by an increase in the temperature of water.

METHODS OF THE COMPOSITION OF SOLUTIONS

If a mol of any soluble compound be dissolved in just enough water to make exactly one litre of solution the result is a volume molar solution. Since the volume of water changes with temperature it is usual to specify that the solution is to be made up to a litre volume at 200 °C. Whenever the word molar is used without qualification this type of solution is meant. Similarly, if half the molar weight, or one-tenth the molar weight of a substance be dissolved in enough water to make exactly one litre of solution the result is a half molar or tenth molar solution, respectively, etc. Gram molar weights of all substances contain the same number of molecules.

Estimates made by various methods show this number to be about 6.06×10^{23} molecules. Hence one litre of any volume molar solution will contain this number of solute molecules, one millilitre (0.001 litre) will contain one thousandth of this number, and so on. Equal volumes of all solutions of the same volume molarity contain the same number of solute molecules but different numbers of solvent molecules. If a given volume of a volume molar solution be diluted with an equal volume of water, the result is a 0.5M solution; if a given volume be diluted with nine volumes of water the result is a 0.1M solution etc. Therefore a volume molar solution of any strength may be diluted with water and the resulting more dilute solution will have a volume molarity in proportion as it has been diluted.If a mol of any soluble substance be completely dissolved in 1000g. ofwater the result is a weight molar solution. Such solutions are often called molal solutions. They are used principally in experimental work upon various osmotic phenonema.The addition of a mol of most solids to a litre of water will increase the volume of the resulting solution to more than one litre. This increase in volume is called the solution volume of the solute. The solution A mol is the molecular weight of a compound in grams.

There is a shrinkage in volume when the solute is added to the solvent. On the other hand the solution volume of some compounds, especially the sugars, is considerable. When a mol of sucrose is added to 1000g. of water the resulting solution will have a volume of 1207cc. at 0 °C. Hence the solution volume of sucrose is 207cc. The solution volume of a mol of sodium chloride, on the other hand, is only about 18 cc. Since every solute has a different solution volume, it follows that equal volumes of weight molar solutions do not contain the same number of either solvent or solute molecules.Neither will the dilution of a given volume of a weight molar solution with an equal volume of water result in a 0.5 molal solution. In other words the concentration of a weight molar solution does not change in proportion to the amount by which it has been diluted. This is a fact which is sometimes overlooked in experimental work.

In physiological work it is often convenient to make up solutions on a percentage basis. Such solutions are made up either on the basis of percentage by weight, or percentage by volume. Solutions of solids in water or other solvents are made up on a weight percentage basis. A 10 per cent sodium chloride solution, for example, is made by dissolving 10 g. of sodium chloride in 90 g. of water. Solutions of liquids in water or other solvents can also be prepared on a weight percentage basis. It is simpler, however, to make up such solutions on a volume percentage basis. On this basis a 10 per cent solution of alcohol is prepared by adding 10cc. of alcohol to 90cc. of water. The percentage system of solutions has no direct relation to either the volume molar or weight molar systems.

RELATIONSHIP BETWEEN ELECTROLYTES AND NON-ELECTROLYTES

Some aqueous solutions readily conduct an electric current; others do not. The former are called electrolytes; the latter non-electrolytes.The solutions of all acids, bases, and salts are electrolytes. Solutions of most organic compounds such as the sugars, alcohols, ketones, and ethers are non-electrolytes.

This process is called electrolysis. If hydrochloric acid is the electrolyte, for example, hydrogen gas will be liberated at the negative pole (cathode) and chlorine gas will be liberated at the positive pole (anode).If electrolysis is continued long enough eventually all of the hydrochloric acid present in the system will be decomposed into hydrogen gas and chlorine gas. Strictly speaking the term electrolyte refers only to a solution of an ionized substance, but it is also often applied to any compound which, when dissolved in water, produces ions.A similar dual usage of the term non-electrolyte also prevails.

The occurrence of electrolysis, as well as the unusually large effects of electrolytes on the osmotic pressures of solutions, have led to an explanation of the behaviour of electrolytes in terms of the theory of electrolytic dissociation. This theory was first advanced by the Swedish chemist Arrhenius who introduced it in essentially its present form in 1887. According to the Arrhenius theory when an electrolyte is dissolved in water some of the molecules dissociate into two kinds of particles, one positively charged, the other negatively charged. Each of these particles is called an ion. Ions are supposed to be present in a solution whether any electrolysis occurs or not.Two, three, four or even more ions may be formed from a single molecule.The conduction of an electrical current by an electrolyte is due to the presence of these ions.

The positively charged ions, which in electrolysis migrate towards the cathode, are called cations; the negatively charged ions which migrate towards the anode are called anions. A cation may carry one, two, three, or even four positive charges; an anion may carry from one to several negative charges.When an electrolyte dissociates the number of positive charges carried on thecations is always equal to the number of negative charges carried by the anions.The Arrhenius theory does not assume complete ionization of all of the solute molecules in an electrolyte but that the molecules are continuously dissociating into ions, while free ions in the solution are continuously reuniting and forming molecules, both processes proceeding at an equal rate whenever an equilibrium condition prevails.An equilibrium of this sort, which is maintained by two opposing processes proceeding at equal rates is termed adynamic equilibrium. Electrolytes vary greatly in degree of dissociation. Those in which a large proportion of the molecules are maintained in a dissociated condition are termed strong electrolytes; those of which the contrary is true, weak electrolytes.Strong electrolyte solutions conduct electric currents better than weak electrolyte solutions of equal concentration. In general the more dilute a given electrolyte solution the larger the proportion of dissociated molecules present. In extremely dilute solutions ionization is practically complete. Increase in temperature reduces the dissociation of an electrolyte.

It is probable that many of the so-called non-electrolytes also dissociate very slightly in solution, but the degree of dissociation of such compounds is so small that it can be detected only by very refined methods, if at all. Even water, as the subsequent discussion will show, dissociates slightly, producing hydrogen and hydroxyl ions.

ACIDS, BASES, AND SALTS

An acid may be defined as a substance which produces hydrogen ions (H^+) when dissolved in water. The characteristic properties of acids are due to the hydrogen ions produced. The most common in Organic acids are hydrochloric (HCl), nitric(HNO_3), and sulfuric($H_{25}O_4$). In living organisms a large group of more complex, but much weaker acids, known as organic acids, play important roles. The strength of an acid depends upon its degree of ionization; the greater the proportion of hydrogen ions an acid produces in solution at a given concentration, the stronger it is.

A base may be defined as a substance which produces hydroxyl ions (OH^-) when dissolved in water. The characteristic properties of bases are due to the hydroxyl ions produced. Some of the commonest bases are sodium hydroxide (NaOH), calcium hydroxide ($Ca(OH)_2$), and potassium hydroxide (KOH). The strength of a base, like that of an acid, depends upon its degree of ionization. The greater the proportion of hydroxyl ions a base produces in solution at a given concentration, the stronger it is.

The theory of electrolytic dissociation accounts much more satisfactorily for the behaviour of weak electrolytes than of strong electrolytes. For this and other reasons Debye and Hiickel in 1923, and other modern investigators, have postulatedthat strong electrolytes, at least, are completely dissociated. The fact that they behave as if only partly dissociated is accounted for in terms of inter ionic attractions. These attractions are supposed to exert what may be roughly described as a braking effect upon the movement of the individual ions.This prevents them from operating at their maximum effectiveness in conducting an electric current or in other phenomena depending upon ionic mobilities. The greater the concentration of an electrolyte, the closer together are the ions, and, according to this view, the less their apparent dissociation.

A salt may be defined as a compound which has been produced by the union of an anion or anions of an acid with the cation or cations of a base. Salts are produced when an acid and a base are brought together in a solution as a result of a chemical union between the hydrogen ion (s) of the acid and the hydroxyl ion (s) of the base, forming water. This reaction is called neutralization.

Since water is practically undissociated, neutralization reactions go rapidly to completion. The reverse reaction, indicated in the above equations by the arrows pointing to the left, is called hydrolysis. Under certain conditions hydrolytic reactions may proceed at a rapid rate, although in the examples presented above the speed of the reaction towards the left (hydrolysis) is negligible compared with the speed of the reaction towards the right (neutralization).

Normal Solutions

The concentrations of acids and bases are most commonly expressed in terms of normal solutions. A normal solution of an electrolyte contains in a dissolved state per litre of solution at 200 C. a weight of the compound in grams equal to its molar weight divided by its hydrogen equivalent. The hydrogen equivalent of a compound is defined as the number of replaceable hydrogen atoms in one molecule, or the number of atoms of hydrogen with which one molecule could react. Thus a normal solution of an acid contains 1.008 g. of replaceable hydrogen per litre; a normal solution of a base 17.008 g. of replaceable hydroxyl per litre. By this system the concentration of any acid, base, or salt can be designated as 0.1 N, 0.5 N, 2 N etc., as the case may be. The normality of an acid solution is a measure of its total acidity, i. e.of its concentration in terms of replaceable hydrogen ions. Similarly the normality of a base solution is a measure of its total basicity. Since 1.008 g.of replaceable hydrogen represents the same number of ions as 17.008 g. of replaceable hydroxyl (why?) it is evident that equal volumes of acid and base solutions of equal normality will exactly neutralize each other. Hydrogen Ion Concentration. The effects of acids upon chemical reactions and upon physico-chemical conditions generally in both inorganic systems and in living organisms are due principally to the hydrogen ions which they produce when in solution.

Some of the most fundamental of physiological phenomena are markedly influenced by the concentration of hydrogen ions in the medium in which they occur. For many purposes therefore it is more important to have some sort of a measuring stick of the concentration of hydrogen ions present in a solution than of the total acidity of the solution.Total acidity, as already noted, is customarily expressed in terms of normality. It is also entirely possible to use the normal system for expressing the concentration of hydrogen ions. In a normal solution of hydrochloric acid, for example, about 78 per cent of the molecules are dissociated.

The term normal as used in the preceding sentence refers to total acidity, that is, to all the ionizable hydrogen whether actually present as ions or combined with anions in the form of molecules. In terms of the hydrogen ions present, however, such a solution is only 0.78 N. The term normal as used in this latter sense refers only to the ionized hydrogen. We may speak therefore of a normal solution of hydrogen ions as well as of a normal solution of an acid.Since no acid is ever completely dissociated, a normal solution of any acid will always be less than normal when its concentration is expressed in terms of the hydrogen ions present.

In order to prepare a normal solution of hydrogen ions it is necessary to make up a solution which is more than normal in terms of total acidity. Such a solution must be of the precise strength and degree of dissociation that exactly 0.008 g. of the ionizable hydrogen present are actually in the dissociated form as ions per litre of the solution. Although hydrogen ion concentration can be readily expressed in terms of normalities, in actual practice this system is not generally used because it is apt to prove cumbersome, especially when it is necessary to refer to the very small concentrations of hydrogen ions usually dealt with in biological problems.

The hydrogen ion concentration of a solution is now quite generally defined in terms of its pH value. The pH value bears a simple mathematical relation to the hydrogen ion concentration of a solution in terms of its normality. Because of the practically universal acceptance of this system it is necessary to understand the significance of the term pH and its relation to hydrogen ion concentration expressed in terms of normality. The relation between pH and hydrogen ion concentration in terms of normality is a logarithmic one. The pH of a normal solution of hydrogen ions is 0, of a 0.1 N solution i, of a 0.01 N solution 2, etc. Zero is the logarithm of i, i is the logarithm of 10, 2 is the logarithm of 100, etc. The pH value for any solution is the negative of the logarithm of the hydrogen ion concentration in terms of normality. It may also be defined as the logarithm of the number of litres of solution that contains one gram atomic weight of hydrogen ions.

All aqueous solutions as well as pure water contain hydrogen ions in some concentration. Corresponding to the concentration of hydrogen ions is a certain definite concentration of hydroxyl ions. It can be shown by the principle of mass action that the mathematical product of the concentration of hydroxyl ions and the concentration of hydrogen ions in a solution is a constant.

This may be expressed as follows:

$$(H^{+}) \times (OH)K.Kio14 \text{ at } 20\ ^{\circ}C.$$

Hence as the pH of a solution is increased the pOH decreases and vice versa. For example, if the pH increases from 5 to 6 the poll decreases from 9 to 8. Furthermore, only at pH 7 can the concentration of hydrogen ions equal the concentration of hydroxyl ions. This is therefore the neutral point on the pH scale and corresponds to the dissociation of pure water. This pH value represents a dissociation of only one water molecule in approximately every 555,000,000.

Values below 7 on the pH scale represent the acid range, those above 7 the alkaline range of the scale. An acid solution is one with a larger concentration of hydrogen ions than hydroxyl ions, while in an alkaline solution the reverse is true. The lower the pH value the greater the hydrogen ion concentration of a solution. A pH value of 5 represents ten times the hydrogen ion concentration of a solution with a pH of 6 and one hundred times the hydrogen ion concentration of one with a ph value of 7, etc. This is due to the fact, previously emphasized, that the numbers on the pH scale are related to each other as logarithms and not as ordinary arithmetic numbers. The hydroxyl ion concentration of a solution could be expressed in pOH units instead of in pH units. For alkaline solutions especially this would seemto be a logical practice. But because of the definite mathematical relationship between the pH and pOH the pH value alone also defines the pOH value. Hence both the acidity of a solution in terms of H + ions, and its alkalinity in terms of OH ions may be, and customarily are, expressed in terms of pH units. It is possible for aqueous solutions to exist which have a pH value of less than 0 (*i.e.* a minus pH value) or more than 14.

An solution of hydrochloric acid, for example, would have a pH value of a little less than o; correspondingly a N solution of sodium hydroxide would have a pH value a little greater than 14. In actual experience, however, solutions of such extreme concentrations are seldom encountered that the sodium acetate solution retards in some way changes in pH value upon the addition of an acid while a solution of sodium chloride does not.

Solutions of such compounds as sodium acetate which are relatively resistant to changes in pH due to the addition or loss of hydrogen or hydroxyl ions are known as buffer solutions, and this property of solutions is called buffer action. Solutions such as sodium chloride which show no buffer effect are called unbuffered solutions.If 1 cc. of a normal sodium hydroxide solution be added to

10 cc. of a normal solution of sodium chloride, a marked increase in pH will occur. Ifi cc. of the same solution be added to 10 cc. of a normal solution of sodium acetate, its pH will also increase markedly. In other words the sodium acetate solution is buffered against the addition of an acid, but not against the addition of a base. If, on the other hand, 10 cc. of a normal solution of acetic acid be substituted for the sodium acetate it will be found that this solutionis strongly buffered against the introduction of hydroxyl ions into the solution.

A mixture of equal volumes of molar sodium acetate solution and molar acetic acid solution will exhibit buffer action against both acids and bases over a considerable range of the pH scale.Two important points regarding buffer solutions have been brought out by the foregoing discussion. No one solute as a rule will act as a buffer against both acids and bases. Furthermore no buffer solution will exert a buffer effect over the entire range of the pH scale. Different buffer solutions vary greatly in their effectiveness in maintaining pH stability. Some are strongly buffered; others weakly. Some are strongly buffered against acids and weakly buffered against bases; of others the converse is true. The commonest types of buffer systems are those composed of a weak acid plus one of its salts.

The sodium acetate-acetic acid buffer system already described is such a system. Practically all of the buffer solutions of importance in living organisms belong to this group. Buffer action consists essentially in the tying up of free hydrogen or hydroxyl ions nearly as rapidly as they are introduced into the solution in the formation of compounds which are only slightly dissociated. The ensuing change of pH is therefore relatively small in proportion to the volume of acid or base added. As an illustration let us consider once more a solution consisting of both sodium acetate and acetic acid dissolved in water. The sodium acetate is strongly dissociated, but the acetic acid will dissociate only slightly, being a weak acid.Suppose now that a little HCl be added to this solution. This is equivalent to adding H_1 and Cl ions and HCl molecules; the latter, however will dissociate, forming additional ions as rapidly as the H_1 and Cl ions already present are bound up in chemical combination.

H^+ and CH_3COO^- ions cannot exist side by side in the same solution in appreciable concentrations since CH_3COOH is a poorly dissociated compound. Hence the added H^+ ions are almost all tied up in the formation of CH_3COOH. The Cl ions form NaCl with the Na ions which dissociates in the usual way. The result is that there is only a slight increase in the concentration of hydrogen ions in the solution, and hence only a very slight reduction in pH value. Suppose now that instead of HCl, a little NaOH be added to this solution.

This is equivalent to adding Na^+ and OH ions and NaOH molecules; the latter, however, will produce additional ions by dissociating as rapidly as the Na^+ and OH ions already present are tied up in chemical combination. But OH and H+ ions cannot exist side by side in the same solution in appreciable concentrations since H_2O is only slightly dissociated. Most of the added OH ions therefore combine with the H^+ ions produced by the CH_3COOH and form H_2O. More of the CH_3COOH hydrolyzes producing more H+ ions, which in turn unite with more of the OH ions. This continues until practically all of the added OH ions are tied up. The Na^+ ions form CH_3COONa with the CH_3COO ions which dissociates in the usual way. The final result is that there is only a very slight decrease in the concentration of H^+ ions in the system, and hence only a very slight increase in its pH value.

Any mechanism which will act in such a way as to remove hydrogen or hydroxyl ions from a solution may operate as a buffer system. Other types of buffering are known but they are relatively of much less importance in living organisms than the type of chemical mechanism which has just been considered. Hydration of Solutes. Molecules of water adhere to the particles of many solutes. Water thus associated with the particles of a solute is called water of hydration. For example, each molecule of dissolved sucrose apparently has six molecules of water associated with it. This means that if a mol of sucrose (342.2 g.) is dissolved in water there will be bound to the sucrose molecules a total of six mols of water (108.096 g.). Ions also are hydrated. Different species of ions appear to have different numbers of water molecules associated with them.Although exact values have been assigned by several experimenters for the number of molecules of water of hydration for certain ions these values are open to question as the results of different investigations do not agree. It seems certain, however, that both anions and cations are hydrated, and that for some kinds of ions, at least, as many as a hundred or more molecules of water of hydration may be associated with a single ion.

ROLE OF INTERFACIAL PHENOMENA IN PLANT PHYSIOLOGY

There are a number of familiar observations which indicate that the surface layer of water possesses certain distinctive properties which are not exhibited within the body of the liquid. For example, if a perfectly dry, clean, steel needle be carefully laid on the surface of some still water it will float, in spite of the fact that steel is heavier than water. Careful observation shows that the water surface beneath the needle is depressed, forming a tiny liquid cradle, within which the needle is supported. This phenomenon is clearly due to the properties of the surface film of the water, because if the needle is laid on the water at a slight angle, so that this film is punctured, it will immediately sink to the bottom of the vessel.

A number of other familiar phenomena are possible principally because of the distinctive properties of the surface film of water. The ability of many kinds of insects to walk on water, the formation of drops by water under certain conditions the rise of water in capillary tubes, soil, blotting paper, etc., the upward bulging of the water surface in a vessel when it is filled to the very last drop, and many other aspects of the behaviour of water, are due largely or entirely to the properties of its surface layer. The surface layers of other liquids also exhibit distinctive properties, but in no common liquid except mercury are such properties as well marked as in water.

SURFACE TENSION

Every molecule within the body of a liquid, although in rapid motion as a result of its kinetic energy attracts and is attracted by other molecules. The attractive forces which any molecule exerts upon its neighbours may be considered as acting along lines of force which radiate from it in all directions. The pull exerted by any molecule upon surrounding molecules, however, diminishes very rapidly with increasing distance from that particular molecule, reaching a maximum at distances not exceeding one or two molecular diameters. Liquids possess a definite volume and distinct boundaries because the molecules are close enough together to attract each other with forces of considerable magnitude.The important property of cohesion in liquids is due to these forces. The magnitude of the cohesive force between the molecules is greater in water than in most liquids. The molecules of a solid are so closely bound together by intermolecular attractions that it has lost ability to change its form, and possesses not only definite, but fixed boundaries. The molecules of both liquids and solids are kinetically active, but their movements are limited more or less rigidly by cohesive forces between the molecules. This limitation upon the freedom of movement of molecules is much greater in solids than in liquids. In gases the distance between the molecules is relatively so great that the intermolecular attractive forces are of negligible magnitude.

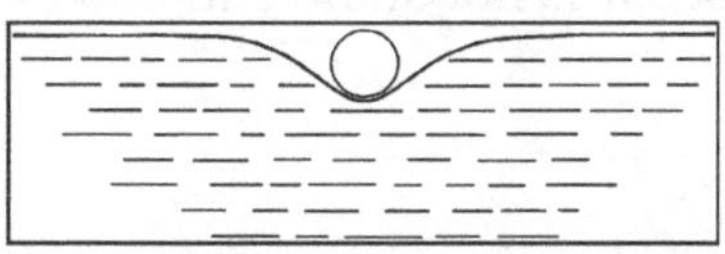

Fig. Needle Floating on water

Molecules at the surface of a liquid or a solid are affected differently bythe cohesive forces between molecules than those within the interior. The further more detailed discussion of this point will be in terms of water. In a vessel of water each of the individual molecules in the body of the liquid is attracted in all directions by the cohesive forces exerted upon it by the surrounding molecules.

At the surface, however, the molecules are not surrounded on all sides by other water molecules. Every molecule in the surface layer is pulled strongly towards the interior by the water molecules beneath it. It is for this reason that drops of water falling freely through space acquire approximately a spherical shape. Since there are no outwardly directed attractions which balance these internally directed forces the surface layer of molecules in any liquid is constantly under a tension. The strained condition of the surface layer of liquid molecules at liquid-gas boundaries is known as surface tension.Solid surfaces also manifest surface tensions, but gases, having enormous intermolecular distances and no surface boundaries, do not. In addition to the unbalanced molecular attractions which the molecules in the surface layer of a liquid sustain, such molecules are usually definitely oriented. This results in a greater concentration of molecules in the surface layer than in the body of the liquid, and contributes towards maintenance of rigidity of the surface film. Surface tension values are generally stated in terms of dynes per centimeter, that is, as the force in dynes which is exerted by the liquid along a line 1 cm. in length.

ISSUES AFFECTING SURFACE TENSION

Temperature

Since the phenomenon of surface tension is one manifestation of the cohesive forces existing among molecules, the greater these cohesive forces, the greater the surface tension. Temperature is a measure of the kinetic energy of the molecules, hence an increase in temperature means an increase in their kinetic energy and vice versa.

As the kinetic activity of water molecules increases,the effectiveness of the cohesive forces existing between them decreases. Hence as the temperature of water increases, its surface tension decreases. The decrease in surface tension of any liquid is, in fact, directly proportional to increase in temperature, except when a liquid approaches its critical temperature. The variation in the surface tension of water within the temperature range at which most physiological processes occur is not sufficient, however to be of any very great significance.

Solutes

The surface tension of a liquid is influenced by the presence of the molecules or ions of a solute. Some solute increase the surface tension of water; others decrease it. If alcohol, for example, be added to water the surface tension of the resulting solution will be less than that of pure water. The attraction between an alcohol molecule and a water molecule is less than that between two water molecules. Since the alcohol molecules are distributed among the water molecules, the mean cohesive force in the solution is less than that in pure water.

The reduction in surface tension occasioned by the addition of alcohol to the water is one aspect of the reduction of the cohesive forces between the molecules in the solution. Most organic compounds have a similar and often very marked effect on the surface tension of water when dissolved in it. However a few such as sucrose have a slight increasing effect on surface tension. Most inorganic salts when dissolved in water increase its surface tensions. The presence of such solutes increases the cohesion within the liquid by attracting the water molecules more strongly than the water molecules attract each other.

The mean cohesive force of the molecules in the solution is thus increased. The result of this effect is a greater surface tension in the solution than in pure water. The magnitude of the effect of inorganic salts upon surface tension is never very great, however.

As a rule, increasing the concentration of a solute in water increases the magnitude of its effect, whether positive or negative, upon surface tension. The influence of solutes upon surface tension is not proportional to their concentration, however, as low concentrations of any solute produce relatively greater effects upon surface tension than high concentrations.

Interfacial Tension

The condition of tension or strain in a limiting layer of molecules is not confined to boundaries between liquids and gases.At a boundary between two immiscible liquids or between a solid and a liquid each of the abutting layers of molecules is subject to similar tensions. In general compounds which decrease the surface tension of water are of the type known as non-polar, while those which increase it are of the type known as polar.

The term surface tension is usually restricted to tensions developed at the surface of a liquid when in contact with a gas, *i.e.* at the interface between a liquid and a gas. Surface tension is therefore merely one variety of interfacial tension. The magnitude of interfacial tensions is influenced by the factors of temperature and solutes in the same general way as surface tensions.

Adsorption

The molecules of most organic compounds, as we have already seen, have a lesser attraction for water molecules than the molecules of water have for each other. This relatively greater attraction between water molecules than between the molecules of water and those of such non-polar solutes results in the displacement into the surface or interfacial layers of a disproportionately large number of the solute molecules. Hence the concentration of most organic solutes is greater in the surface or interfacial layers than in the body of the liquid. The concentrating of solute molecules in the surface or interface of a liquid is called positive adsorption.

The molecules of inorganic salts, on the other hand, are more strongly attracted to water molecules than the water molecules are attracted to each other. Because of this stronger attraction of the water molecules for such polar solutes relatively more solute molecules are pulled from the surface films into the body of the liquid than water molecules. Molecules of polar solutes, therefore, generally occur in greater concentration in the body of the liquid than in the surface layers. This phenomenon is called negative adsorption. When negative adsorption occurs the solvent molecules become relatively more concentrated in the surface layer than in the body of the liquid.For reasons which will become clear later, negative adsorption rarely occurs at interfaces.

The difference in concentration between the surface layer and interior of a liquid are in general much less in negative adsorption than in positive adsorption. Adsorbed molecules are not static, but exhibit a continuous although reduced kinetic activity. Molecules adsorbed at surfaces or interfaces are in dynamic equilibrium with the molecules of the same species in the body of the liquid. An adsorption equilibrium is attained when the number of adsorbed molecules passing out of the interface in a unit of time is equal to the number passing into the interface.When adsorption occurs from a liquid containing many solutes, as is true of most biological liquids, all solutes are adsorbed to a greater or lesser degree, depending upon their specific properties in relation to the adsorbing surface. In general, however, under such conditions no one solute will be adsorbed as completely as if it alone were present. If a substance is introduced into an aqueous solution which will lower the surface tension of water more than another substance, already adsorbed, the first compound will largely displace the second compound in the interfaces of that solution.

Interfacial Adsorption

Adsorption occurs at all types of interfaces.Many solids adsorb gas molecules very powerfully at solid-air interfaces.Solutes may be adsorbed at interfaces between immiscible liquids, and at interfaces between liquids and solids, as well as at liquid-gas interfaces (surfaces). Water molecules are strongly adsorbed at many solid-liquid interfaces. Adsorption is, therefore, a phenomenon of a very general occurrence. Adsorption at liquid-gas interfaces involves only the forces of attraction between the solvent and solute molecules, since the molecules of a gas are too far apart to have any appreciable attraction for the solute molecules. Similarly adsorption at solid-gas interfaces involves only the forces of attraction between the molecules of the solid and those of the adsorbed gas. At liquid-liquid, or solid-liquid interfaces, however, adsorbed molecules are subject to the attraction of the molecules of the two abutting substances at the interface.

At an interface between carbon particles and water, molecules are attracted by both the carbon and the water. If the carbon exerts a greater attraction for the solute than the water as is usually the case a greater accumulation of solute molecules will occur in the interface, than if the opposite is true. The concentration of adsorbed molecules which will be attained at any solid-liquid or liquid-liquid interface is therefore controlled by the relative magnitude of the attractive forces of the two phases for the molecules of the adsorbate (substance adsorbed). Inorganic salts which are negatively adsorbed at surfaces (water-air interfaces) are usually positively adsorbed at water-liquid interfaces, and almost always so at water-solid interfaces. This is due to a greater attraction for the solute particles by the non-waterphase of the interface than the water molecules themselves exert. Specially treated (activated) charcoal is one of the best adsorbents known both for gases and for solutes. Many important applications are made of this property of charcoal in chemistry and technology. Gas masks for protection against poison gases in warfare owe their efficiency to the adsorptive capacity of charcoal. *Colour*ed or other soluble impurities are often removed from liquids by passing them through layers of charcoal.Solids such as charcoal which are powerful adsorbents usually possess an enormous surface area in proportion to their mass, which accounts for their effectiveness as adsorbing agents. The surface of g. of charcoal has been variously estimated at from about 50 to about 600 square metres. Much of this surface is in the form of internal submicroscopic. Adsorption in which the fundamental controlling forces are the cohesive and adhesive forces between molecules is called mechanical adsorption in order to distinguish it from certain more complex kinds of adsorption phenomena. Of these the most important is electrical adsorption. Most surfaces in contact with water bear an electrical charge. The presence of such charges markedly influences the adsorption of ions or other charged particles.

Cellulose, in common with many other substances, acquires a negative charge when immersed in water. If a strip of cellulose filter paper be arranged so that its lower end dips in a solution of eosin (a red dye) it will be observed that the eosin will rise through the filter paper almost as rapidly as the water from the solution rises by capillarity. If another strip of filter paper be similarly arranged to dip in a solution of methylene blue (a blue dye) the water will rise up the filter paper as rapidly as it does from the eosin solution, but the dye will scarcely rise at all. The difference in behaviour of these two dyes is due principally to the electrical charge which their coloured ions carry.

The coloured ions of eosin are negatively charged; those of methylene blue positively charged. The negatively charged particles of eosin are repelled from the negatively charged surface of the cellulose fibres. The eosin particles are therefore driven towards the centre of the capillary columns of water in between the fibres, and move up the filter paper almost as rapidly as the water. As soon as the positively charged particles of methylene blue come in contact with the negative charges of the cellulose, however, they are held to the surface of the cellulose by the forces of electrical attraction. In other words they are electrically adsorbed. While the water rises by capillarity at about the same rate as it does from the eosin solution, the rise of methylene blue is relatively very slow.

Certain examples of adsorption have also been recognized which are designated by the name of chemical adsorption. Chemical reactions are involved in adsorption phenomena of this type. The familiar reaction of 12Kl with starch in which the starch stains purple is often considered to be an example of chemical adsorption. Actually there exists no clean cut distinction between mechanical, electrical, and chemical adsorption. The several phenomena intergrade, and elements of more than one of these mechanisms are usually operating whenever adsorption occurs. Since the surface or interface of even pure water bears an electric charge, at least minor electrical effects are involved in even the simplest adsorption phenomena.

Adsorption of Water

Water itself is strongly adsorbed at certain types of interfaces. The adsorption of water is a phenomenon which can be regarded as more nearly similar to electrical adsorption than to mechanical adsorption. Water molecules, while not charged in the sense that ions are strongly polar and behave somewhat as charged bodies. Water-vapour molecules become adsorbed on many different kinds of surfaces.

The water-vapour adsorbed upon glass weighing bottles often becomes an appreciable source of error in accurate quantitative work. The particles in certain colloidal systems have the property of adsorbing large quantities ofwater, as shown in the next two chapters. A large portion of the water which moves into various types of substances in the process of imbibition becomes adsorbed upon the internal surfaces of the imbibing substance.

Biological impact of Adsorption

Adsorption phenomena are known to play a manifold role in living organisms, probably being involved in practically all cell activities. Protoplasm and many other constituents of plant cells are essentially colloidal, and adsorption phenomena are of general occurrence in colloidal systems. Within plant cells many interfaces occur, as at the boundaries between the proto-plasm and vacuole, protoplasm and cell wall, and nucleus and cytoplasm, at all of which interfacial concentration of solutes undoubtedly occurs. The adsorption of certain compounds at the cytoplasmic interfaces is generally believed to exert a marked influence on the permeability of the cytoplasm. It is quite possible that interfacial tensions play an important role in the process of cell division. Imbibitional phenomena, of basic importance in the water relations of plant cells, involve the adsorption of water. The action of enzymes as well as of other catalysts is generally believed to involve adsorption phenomena. Much information regarding the structure of cells has been gained by the use of dyes which are differentially adsorbed by various constituents of cells. Chromosomes, for example, are so named because they strongly adsorb certain stains.

10

Diffusion of Gases

If a small vial of bromine gas is broken under a bell jar which has previously been evacuated of air the entire jar quickly becomes filled with the brownish vapour of bromine. The distribution of the bromine gas throughout the bell jar has been accomplished by the kinetic activity of the bromine molecules, and is a simple example of the process of diffusion. If the vial of bromine be broken under a bell jar which has not been evacuated of air, the time required for the bromine gas to completely occupy the bell jar will be longer than when diffusion of the gas occurs into a vacuum. Under such conditions the freedom of movement of the bromine molecules is greatly impeded by the presence of molecules of the gases of the air, and the diffusion process is retarded. If the pressure of the air within the bell jar be increased to two atmospheres (which is equivalent to doubling the concentration of all the gases in the jar), the rate of diffusion of the bromine gas through the air would be still less than when the jar was occupied by air at atmospheric pressure.

Many other simple examples of the diffusion of gases might be cited. If a bottle of ammonia, ether, peppermint oil, or of any other readily volatile substance with a characteristic odour be opened indoors, within a very short time the distinctive odour of that substance can be detected in all parts ofthe room. Such a dispersal of gas molecules is accomplished at least in part by diffusion, although air currents often assist in speeding up such a distribution of molecules. Except in the rare case of diffusion into a vacuum diffusing molecules move between the molecules of other substances. The following some what fanciful analogy may aid in a visualization of the kinetics of the diffusion process in gases. Suppose two adjoining rooms to be connected by a closed double door. Imagine also that one of these rooms contains a large number of tennis balls travelling in various directions along straight pathways at different rates of speed.

The average distances between the tennis balls are supposed to be relatively great in proportion to their diameters. Each tennis ball represents a molecule. The individual balls will be constantly bumping into each other and into the walls of the room. Because of the large number of balls present innumerable collisions will occur every second. Each time a ball strikes a wall of the room it will bounce off along a different linear pathway. Similarly, whenever two balls collide, each will be deflected out of its course along a different route, to which it will hold undeviatingly until it is again deflected from its path by another collision.

The course of each ball will thus be a zigzag progression through space, each short segment of its path being terminated by a collision which changes its direction. All of the balls do not move at the same speed at any given time but the speed of each fluctuates from moment to moment as a result of the numerous collisions in which it participates. The average speed of the entire group will remain constant, however, as long as the temperature remains unchanged. As a result of this haphazard mutual buffeting, the tennis balls will remain essentially equally distributed throughout the room.

Suppose now that the double doors connecting the two rooms are thrown open. As a result of their haphazard movement some of the balls close to the door will pass into the empty room. The first ones to do this will travel without interruption until they bump into one of the walls, as their rate of progression will not be impeded by collisions with other molecules. When they hit against one of the walls of the room they will bounce back into its interior along a new pathway. As more and more of the balls pass into the originally empty room their rate of progress becomes slower since the greater their concentration in the room, the greater the number of collisions per unit of time.

As soon as any appreciable number of tennis balls has invaded the empty room, as a result of their random movements, some will pass back through the doorway into the room which originally contained all of them. As long, however, as the concentration (number per unit volume) of the tennis balls is greater in one room than in the other their random movements will result in more passing through the door into the room in which their lesser concentration prevails, than in the opposite direction.In a relatively short time the concentration of tennis balls will have become equal in both rooms. In other words they have diffused from one room into the other.

After equality of concentration has been established the number of balls passing through the door in one direction in any interval of time will be exactly equal to the number moving in the opposite direction. When this condition of dynamic equilibrium is attained, diffusion, in the sense the word will be used in this discussion, is no longer occurring. If, in the hypothetical illustration just described, one room were filled with white tennis balls, and the other with red tennis balls, diffusion would occur simultaneously in both directions. The red balls would diffuse towards the room in which their initial concentration was zero while the white balls would diffuse in the opposite direction. At equilibrium the concentration of the white balls would be equal throughout the two rooms, and this would likewise be true of the red balls.

Even after a dynamic equilibrium has been attained in any system haphazard kinetic activity of the molecules continues. This is sometimes referred to as diffusion, but will not be so considered in this book. The term diffusion will be used only to characterize situations in which there is gain in the number of molecules of a certain kind in one part of a system at the expense of other parts. According to this concept diffusion can only occur when the concentration of the diffusing substance is not uniform throughout the system, and the process can continue only as long as differences of concentration are maintained. Diffusion dependent upon the kinetic energy of molecular motion and thus resulting solely from differences in concentration is often called simple diffusion in order to distinguish it from more complex types of diffusion phenomena in which other forces come into play. Examples of such more complex phenomena will be encountered in some of the following chapters. The phenomenon of diffusion is exhibited by the molecules of liquids solutes, and even solids, as well as by those of gases. Diffusion of ions also occurs, and even colloidal particles diffuse, but only at very slow rates.

Diffusion phenomena should be clearly distinguished from mass movement in which the moving units are not single molecules, but more or less extensive assemblages of molecules. Winds and air currents generally are examples of mass movements of gas molecules. The draft of warm air ascending a chimney is another. All of the phenomena just listed, and many others, are due primarily to differences in the density of the gases in various parts of a system. Heavy gases are more strongly attracted by gravitational forces than light gases. Hence, as in a chimney, cold (relatively heavy) air, displace shot (relatively light) air, forcing the latter to rise. Similar phenomena on a grand scale are the principal cause of winds and air currents. They are representative of the physical process called convection. Such phenomena also occur in liquids. Mass movements of gases and liquids can also be caused in many other ways.

DIFFUSION PRESSURE

That diffusing gases sometimes result in the development of measurable pressures can be readily shown by certain simple experiments. This apparatus consists essentially of a vertically arranged glass tube, to the upper end of which is attached, by means of rubber stopper, a cylindrically shaped porous clay cup. This cylinder is hollow and its thick walls are pierced by numerous minute capillaries. These pores are fine enough to prevent mass movement of gases at any appreciable rate, but the porous clay is not a membrane in the usual sense of the word. The lower end of the vertical glass tube dips in some coloured water. If such a porous clay cylinder, containing air, be enclosed within a bottle containing hydrogen gas a rapid bubbling of gas will occur from the lower end of the tube through the dye solution into which it dips. When the bottle is removed a rapid and sudden rise of liquid up the glass tube will ensue. This is followed by a slow subsidence in the level of the liquid in the glass tube, until it falls to that of the water in the beaker.The explanation of this sequence of events is as follows. For reasons which are discussed later the rate of diffusion of hydrogen is greater than that of nitrogen or oxygen under comparable conditions. When the porous clay cup is first enclosed within the bottle of hydrogen, rapid diffusion of hydrogen gas occurs through the pores of the cup. This raises the total gas pressure within the cup, since outward diffusion of oxygen and nitrogen occurs at a much slower rate. Hydrogen gas diffuses into the cup in spite of the fact that this results in a greater total pressure inside the cup than in the surrounding atmosphere.

The direction of the diffusion of the hydrogen gas is controlled entirely by its own differences in concentration, and is unaffected by the presence of other gases. The greater gas pressure inside the cup results in the outward bubbling of a mixture of all three gases at the lower end of the vertical glass tube. When the bottle is removed from around the porous clay cup, the direction of the diffusion of hydrogen is reversed, since the concentration of hydrogen gas is now greater inside of the cup. Since, while enclosed in the bottle of hydrogen, some of the nitrogen and oxygen diffused out of the cup and some was lost by the bubbling of gases from the lower end of the glass tube, this rapid outward diffusion of hydrogen results temporarily in a lower total gas pressure inside the cup than that originally present. Hence the solution rises rapidly in the glass tube. Finally, nitrogen and oxygen diffuse slowly inward, because of the reduced concentration of these gases inside of the cup, and the liquid slowly falls in the glass tube to its original level. That the process of diffusion may result in the development of pressure also can be shown very strikingly by means of another simple experiment. If a pure rubber balloon containing only a little air be suspended in a closed bottle of carbon dioxide gas it will gradually become distended. Rubber membranes are quite readily permeable to the molecules of carbon dioxide but virtually impermeable to those of oxygen, nitrogen, or other gases of the atmosphere.

Since, temperature remaining constant, the pressure exerted by any gas is directly proportional to its concentration (number of molecules per unit volume), it follows that the diffusion of the molecules of gases may be interpreted in terms of the differences in the partial pressure exerted by that gas in different parts of a system. Because of the greater partial pressure of carbon dioxide in the surrounding atmosphere, diffusion of this gas through the walls of the balloon continues until the partial pressure of the carbon dioxide is the same on the two sides of the rubber membrane. Since the initial partial pressure of carbon dioxide inside the balloon was virtually zero, and that in the bottle equivalent to one atmosphere, the inward diffusion of carbon dioxide gas results in a considerable distension of the balloon. In the analysis of diffusion phenomena it is often clarifying to speak of the partial pressure of a gas as its diffusion pressure. Solutes and liquids may also be considered to possess a diffusion pressure, although the existence of such a physical quantity is less evident in such systems than in gases. Hence the concept that diffusion is the movement of the molecules of a substance from a region of its greater to a region of its lower diffusion pressure K often more satisfactory than the interpretation of diffusion phenomena in terms of concentration differences.

PRINCIPLE OF INDEPENDENT DIFFUSION

In the first experiment described in the preceding section it was shown that while one gas (hydrogen) was diffusing in an inward direction through the pores of a clay cylinder, other gases (oxygen and nitrogen) were simultaneously diffusing in an outward direction through the same pores. This exemplifies one of the most important principles governing diffusion phenomena, namely that the direction in which any substance will diffuse is controlled entirely by its own differences in diffusion pressure, and is not influenced by either the direction or rate of diffusion of other substances in the same system.

Hence, in any given system, as for example, two adjacent plant cells, a number of substances may be diffusing in one direction across the intervening membranes, while simulta-neously other compounds may be diffusing in the opposite direction across the same membranes. Each one of these individual substances will diffuse in the direction determined by its own differences in diffusion pressure, and at a speed which is determined by the factors which are influencing the diffusion of that particular substance.

FACTORS INFLUENCING THE RATE OF DIFFUSION OF GASES

DENSITY OF THE GAS

Different gases diffuse at different rates even when influenced by the same set of environmental factors. Hydrogen, for example, diffuses more rapidly than any other gas. The same Thomas Graham who conducted some of the earliest studies on the properties of colloidal systems also was one of the first investigators to study quantitatively the phenomenon of gaseous diffusion. He discovered the principle, often called Grahams Law of Diffusion, that the relative speeds of diffusion of different gases are inversely proportional to the square roots of their relative densities. By relative density is meant the weight of a given volume of gas as compared with the weight of the same volume of hydrogen. The relative density of oxygen is 16.

Hydrogen gas will therefore diffuse four times as rapidly as oxygen gas under the same conditions of temperature and pressure. The relative density of carbon dioxide is 22; hence by similar reasoning it is apparent that hydrogen gas will diffuse nearly five times as rapidly as carbondioxide gas. Since a molar weight of any gas occupies a volume of 22.4 litres at standard conditions, molar weights of gases are in themselves a measure of the relative density of gases.

Since the molar weight of hydrogen (H_2) is 2.016, the relative density of any gas on the basis H i, is equal to its molar weight divided by 2.016 (usually rounded off to 2).

TERNPERATURE

Increase in temperature increases the speed of diffusion. This is due, at least in part, to the correlated increase in the kinetic activity of the molecules of the diffusing substance. Actual measurements of the temperature coefficient 2 of diffusion generally yield values between 1.2 and 1.3. Such values are characteristic of any purely physical process such as diffusion.

DIFFUSION GRADIENT

The fact that the direction of the diffusion of gases is from a region of their greater diffusion pressure to a region of their lesser diffusion pressure has already been emphasized. The speed of diffusion is also influenced by differences in diffusion pressure. In general the greater the difference in diffusion pressures between the two regions, the more rapidly diffusion will occur. The rate of diffusion is influenced, however, not only by the difference in diffusion pressures but also by the distance through which the diffusing molecules must travel. These two factors are components of what may be called the diffusion pressure gradient or concentration gradient. In both parts, A and B of this diagram, oxygen is represented as diffusing from a region in which its diffusion pressure is maintained at 760 mm.

Hg (atmospheric pressure) into a region in which its diffusion pressure is just half as great. The length of the connecting tube is, however just twice as great in A as in B. Under the conditions postulated the rate of diffusion from the region of greater diffusion pressure to the other will be just twice as rapid in B as in A. The diffusion pressure gradient is equivalent to the difference in the diffusion pressures between the delivering and receiving ends of the diffusion system divided by the length of the distance between. The greater or steeper this gradient the more rapidly diffusion will occur. The steeper the gradient, the more rapid the change in diffusion pressure per unit of length along the axis of the diffusion gradient. The steepness Of a diffusion pressure gradient may be changed by varying either factor, the difference in diffusion pressures, or the length of the gradient. The temperature coefficient of any process, physical, chemical, or physiological, is defined as the number of times that the rate of the process increases with a 200 °C. rise in temperature. If the rate of the process is doubled the temperature coefficient is 2, etc.

CONCENTRATION OF THE MEDIURN THROUGH WHICH DIFFUSION OCCURS

In general the more concentrated the medium, *i.e.* the more molecules per unit volume in the medium through which the diffusing molecules must pass, the slower the rate of diffusion. Bromine gas diffuses more rapidly through a vacuum than through air.

DIFFUSION OF SOLUTES

The molecules or ions of a solute possess sufficient kinetic energy to move from place to place within the limits of a solution. The simplest method of demonstrating the diffusion of a solute is to introduce a crystal of copper sulphate, or some other compound which is coloured when in solution, into the bottom of a tall glass cylinder filled with water. The cylinder should then be placed in an environment of equable temperature where it will be free from disturbance.The diffusion of the molecules or the ions which pass into solution in the water can be followed by the slow change in colour of the water. One of the most striking facts illustrated in such experiments is the extremely slow rate of diffusion of solutes through water. This is partly due to the fact that in such an experiment the steepness of the diffusion gradient decreases with time, but principally to the fact that the densely packed molecules of the liquid enormously impede the diffusion of the dissolved molecules or ions. The true rate of diffusion of solutes is probably even less than the rates indicated in such experiments, since in all such setups convection currents may develop in the water and aid in the distribution of the solute throughout the body of the liquid. The diffusion of solutes is often demonstrated by employing a gel rather than a liquid as the medium into which diffusion occurs. Such a technique avoids errors introduced by the development of convection currents.

The direction of the diffusion of any solute occurs in accordance with its own differences in diffusion pressure, regardless of the rate or direction of diffusion of other solutes in the same system. The rate of diffusion of solute particles is governed by principles essentially similar to those which control the rate of diffusion of gases and is controlled by the following factors:

SIZE AND MASS OF THE DIFFUSING PARTICLE

Small molecules or ions diffuse more rapidly than large ones. A hydrogen ion, for example, diffuses many times more rapidly than a glucose molecule. Similarly, highly hydrated ions diffuse more slowly than those which have fewer water molecules bound to them, since the association of water of hydration with a molecule or ion in effect increases its size. The mass of the particle will also be a factor influencing the speed of its diffusion. As between two particles of the same size, but different masses, the heavier particle will diffuse more slowly.

TEMPERATURE

The kinetic activity, and hence the rate of diffusion of solute molecules, increases with increase in temperature.

DIFFUSION GRADIENT

The steeper the diffusion gradient, the more rapidly solute particles diffuse.

SOLUBILITY

In general, the more soluble a substance is in a liquid the more rapidly it will diffuse through that liquid. This influence of solubility upon diffusion rates can be interpreted principally in terms of its effect upon the diffusion gradient, since obviously steeper gradients can be built up if the solute is very soluble in the liquid than if it is only slightly soluble.

11

Photosynthesis

Sunlight plays a much larger role in our sustenance than we may expect: all the food we eat and all the fossil fuel we use is a product of photosynthesis, which is the process that converts energy in sunlight to chemical forms of energy that can be used by biological systems. Photosynthesis is carried out by many different organisms, ranging from plants to bacteria. The best known form of photosynthesis is the one carried out by higher plants and algae, as well as by cyanobacteria and their relatives, which are responsible for a major part of photosynthesis in oceans.

All these organisms convert CO_2 (carbon dioxide) to organic material by reducing this gas to carbohydrates in a rather complex set of reactions. Electrons for this reduction reaction ultimately come from water, which is then converted to oxygen and protons. Energy for this process is provided by light, which is absorbed by pigments (primarily chlorophylls and carotenoids).

Chlorophylls absorb blue and red light and carotenoids absorb blue-green light, but green and yellow light are not effectively absorbed by photosynthetic pigments in plants; therefore, light of these colors is either reflected by leaves or passes through the leaves. This is why plants are green.

Other photosynthetic organisms, such as cyanobacteria (formerly known as blue-green algae) and red algae, have additional pigments called phycobilins that are red or blue and that absorb the colors of visible light that are not effectively absorbed by chlorophyll and carotenoids.

Yet other organisms, such as the purple and green bacteria (which, by the way, look fairly brown under many growth conditions), contain bacteriochlorophyll that absorbs in the infrared, in addition to in the blue part of the spectrum. These bacteria do not evolve oxygen, but perform photosynthesis under anaerobic (oxygen-less) conditions.

These bacteria efficiently use infrared light for photosynthesis. Infrared is light with wavelengths above 700 nm that cannot be seen by the human eye; some bacterial species can use infrared light with wavelengths of up to 1000 nm. However, most pigments are not very effective in absorbing ultraviolet light (<400 nm), which also cannot be seen by the human eye. Light with wavelengths below 330 nm becomes increasingly damaging to cells, but virtually all light at these short wavelengths is filtered out by the atmosphere (most prominently the ozone layer) before reaching the earth.

Even though most plants are capable of producing compounds that absorb ultraviolet light, an increased exposure to light around 300 nm has detrimental effects on plant productivity.

PROCESS OF PHOTOSYNTHESIS

Photosynthesis is a chemical process that converts carbon dioxide into organic compounds, especially sugars, using the energy from sunlight. Photosynthesis occurs in plants, algae, and many species of bacteria, but not in archaea. Photosy nthetic organisms are called *photoautotrophs*, since they can create their own food. In plants, algae, and cyanobacteria, photosynthesis uses carbon dioxide and water, releasing oxygen as a waste product.

Photosynthesis is vital for all aerobic life on Earth. In addition to maintaining normal levels of oxygen in the atmosphere, photosynthesis is the source of energy for nearly all life on earth, either directly, through primary production, or indirectly, as the ultimate source of the energy in their food, the exceptions being chemoautotrophs that live in rocks or around deep sea hydrothermal vents.

The rate of energy capture by photosynthesis is immense, approximately 100 terawatts, which is about six times larger than the power consumption of human civilization. As well as energy, photosynthesis is also the source of the carbon in all the organic compounds within organisms' bodies. In all, photosynthetic organisms convert around 100–115 petagrams of carbon into biomass per year.

Although photosynthesis can happen in different ways in different species, some features are always the same. For example, the process always begins when energy from light is absorbed by proteins called photosynthetic reaction centers that contain chlorophylls. In plants, these proteins are held inside organelles called chloroplasts, while in bacteria they are embedded in the plasma membrane. Some of the light energy gathered by chlorophylls is stored in the form of adenosine triphosphate (ATP).

The rest of the energy is used to remove electrons from a substance such as water. These electrons are then used in the reactions that turn carbon dioxide into organic compounds. In plants, algae and cyanobacteria, this is done by a sequence of reactions called the Calvin cycle, but different sets of reactions are found in some bacteria, such as the reverse Krebs cycle in *Chlorobium*. Many photosynthetic organisms have adaptations that concentrate or store carbon dioxide. This helps reduce a wasteful process called photorespiration that can consume part of the sugar produced during photosynthesis.

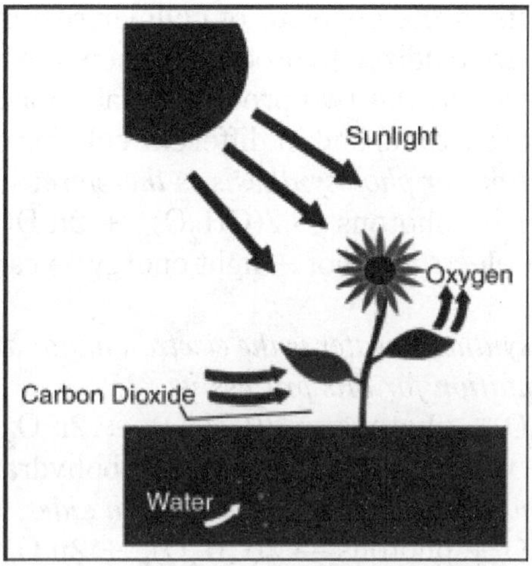

Fig. Schematic of Photosynthesis in Plants.

The first photosynthetic organisms probably evolved about 3,500 million years ago, early in the evolutionary history of life, when all forms of life on Earth were microorganisms and the atmosphere had much more carbon dioxide. They most likely used hydrogen or hydrogen sulfide as sources of electrons, rather than water. Cyanobacteria appeared later, around 3,000 million years ago, and drastically changed the Earth when they began to oxygenate the atmosphere, beginning about 2,400 million years ago.

This new atmosphere allowed the evolution of complex life such as protists. Eventually, no later than a billion years ago, one of these protists formed a symbiotic relationship with a cyanobac terium, producing the ancestor of many plants and algae. The chloroplasts in modern plants are the descendants of these ancient symbiotic cyanobacteria.

Photosynthetic organisms are photoautotrophs, which means that they are repositories of energy, they are able to synthesize food directly from carbon dioxide, water, and using energy from light. They accrue it as part of their potential energy. However, not all organisms that use light as a source of energy carry out photosynthesis, since *photoheterotrophs* use organic compounds, rather than carbon dioxide, as a source of carbon. In plants, algae and cyanobacteria, photosynthesis releases oxygen. This is called *oxygenic photosynthesis*. Although there are some differences between oxygenic photosynthesis in plants, algae and cyanobacteria, the overall process is quite similar in these organisms. However, there are some types of bacteria that carry out anoxygenic photosynthesis, which consumes carbon dioxide but does not release oxygen.

Carbon dioxide is converted into sugars in a process called carbon fixation. Carbon fixation is a redox reaction, so photosynthesis needs to supply both a source of energy to drive this process, and the electrons needed to convert carbon dioxide into a carbohydrate, which is a reduction reaction. In general outline, photosynthesis is the opposite of cellular respiration, where glucose and other compounds are oxidized to produce carbon dioxide, water, and release chemical energy. However, the two processes take place through a different sequence of chemical reactions and in different cellular compartments.

The general equation for photosynthesis is therefore:

$2n\ CO_2 + 2n\ DH_2 + \text{photons} \rightarrow 2(CH_2O)_n + 2n\ DO$.

Carbon dioxide + electron donor + light energy → carbohydrate + oxidized electron donor.

In oxygenic photosynthesis water is the electron donor and, since its hydrolysis releases oxygen, the equation for this process is:

$2n\ CO_2 + 4n\ H_2O + \text{photons} \rightarrow 2(CH_2O)_n + 2n\ O_2 + 2n\ H_2O$.

carbon dioxide + water + light energy → carbohydrate + oxygen + water.

Often 2n water molecules are cancelled on both sides, yielding:

$2n\ CO_2 + 2n\ H_2O + \text{photons} \rightarrow 2(CH_2O)_n + 2n\ O_2$.

carbon dioxide + water + light energy ! carbohydrate + oxygen.

Other processes substitute other compounds (such as arsenite) for water in the electron-supply role; the microbes use sunlight to oxidize arsenite to arsenate:

The equation for this reaction is:

$CO_2 + (AsO_3^{3-}) + \text{photons} \rightarrow (AsO_4^{3-}) + CO$.

carbon dioxide + arsenite + light energy → arsenate + carbon monoxide (used to build other compounds in subsequent reactions).

Photosynthesis occurs in two stages. In the first stage, *light-dependent reactions* or *light reactions* capture the energy of light and use it to make the energy-storage molecules ATP and NADPH. During the second stage, the *light-independent reactions* use these products to capture and reduce carbon dioxide.Most organisms that utilize photosynthesis to produce oxygen use visible light to do so, although at least three use infrared radiation.

REACTION CENTERS AND ANTENNAE

Photosynthetic pigments come in a huge variety: there are many different types of (bacterio) chlorophyll, carotenoids, and phycobilins, differing from each other in their precise chemical structure. Pigments generally are bound to proteins, which provide the pigment molecules with the appropriate orientation and positioning with respect to each other. Light energy is absorbed by individual pigments, but is not used immediately by these pigments for energy conversion. Instead, the light energy is transferred to chlorophylls that are in a special protein environment where the actual energy conversion event occurs: the light energy is used to transfer an electron to a neighboring pigment.

Pigments and protein involved with this actual primary electron transfer event together are called the reaction center. A large number of pigment molecules (100-5000), collectively referred to as antenna, "harvest" light and transfer the light energy to the same reaction center. The purpose is to maintain a high rate of electron transfer in the reaction center, even at lower light intensities.Many antenna pigments transfer their light energy to a single reaction center by having this energy "hop" to another antenna pigment, and yet to another, etc., until the energy is "trapped" in the reaction center. Each step of this energy transfer must be very efficient to avoid a large loss in the overall transfer process, and the association of the various pigments with proteins ensures that transfer efficiencies are high by having appropriate pigments close to each other, and by having an appropriate molecular geometry of the pigments with respect to each other. An exception to the rule of protein-bound pigments are green bacteria with very large antenna systems: a large part of these antenna systems consists of a "bag" (named chlorosome) of up to several thousand bacteriochlorophyll molecules that interact with each other and that are not in direct contact with protein.

In many systems the size of the photosynthetic antenna is flexible, and photosynthetic organisms growing at low light (in the shade, for example) generally will have a larger number of antenna pigments per reaction center than those growing at higher light intensity. However, at high light intensities (for example, in full sunlight) the amount of light that is absorbed by plants exceeds the capacity of electron transfer initiated by reaction centers.

Plants have developed means to convert some of the absorbed light energy to heat rather than to use the absorbed light necessarily for photosynthesis. However, in particular the first part of photosynthetic electron transfer in plants is rather sensitive to overly high rates of electron transfer, and part of the photosynthetic electron transport chain may be shut down when the light intensity is too high; this phenomenon is known as photoinhibition.

Photosynthetic Electron Transfer

The initial electron transfer (charge separation) reaction in the photosynthetic reaction center sets into motion a long series of redox (reduction-oxidation) reactions, passing the electron along a chain of cofactors and filling up the "electron hole" on the chlorophyll, much like in a bucket brigade. All photosynthetic organisms that produce oxygen have two types of reaction centers, named photosystem II and photosystem I (PS II and PS I, for short), both of which are pigment/protein complexes that are located in specialized membranes called thylakoids. In eukaryotes (plants and algae), these thylakoids are located in chloroplasts (organelles in plant cells) and often are found in membrane stacks (grana). Prokaryotes (bacteria) do not have chloroplasts or other organelles, and photosynthetic pigment-protein complexes either are in the membrane around the cytoplasm or in invaginations thereof (as is found, for example, in purple bacteria), or are in thylakoid membranes that form much more complex structures within the cell (as is the case for most cyanobacteria).

All chlorophyll in oxygenic organisms is located in thylakoids, and is associated with PS II, PS I, or with antenna proteins feeding energy into these photosystems. PS II is the complex where water splitting and oxygen evolution occurs. Upon oxidation of the reaction center chlorophyll in PS II, an electron is pulled from a nearby amino acid (tyrosine) which is part of the surrounding protein, which in turn gets an electron from the water-splitting complex. From the PS II reaction center, electrons flow to free electron carrying molecules (plastoquinone) in the thylakoid membrane, and from there to another membrane-protein complex, the cytochrome b_6f complex. The other photosystem, PS I, also catalyzes light-induced charge separation in a fashion basically similar to PS II: light is harvested by an antenna, and light energy is transferred to a reaction center chlorophyll, where light-induced charge separation is initiated. However, in PS I electrons are transfered eventually to NADP (nicotinamid adenosine dinucleotide phosphate), the reduced form of which can be used for carbon fixation.

The oxidized reaction center chlorophyll eventually receives another electron from the cytochrome b_6f complex. Therefore, electron transfer through PS II and PS I results in water oxidation (producing oxygen) and NADP reduction, with the energy for this process provided by light (2 quanta for each electron transported through the whole chain). A schematic overview of these processes is provided in Figure shown below.

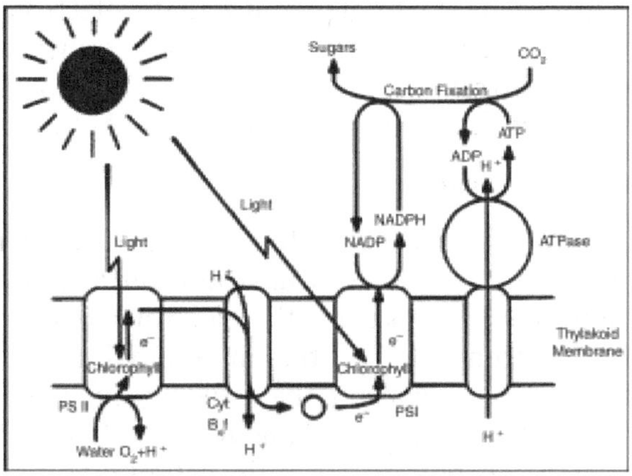

Fig. Overview of Photosynthetic Processes as they Occur in Plants, Algaev, and Cyanobacteria.

CARBON FIXATION

Electron flow from water to NADP requires light and is coupled to generation of a proton gradient across the thylakoid membrane. This proton gradient is used for synthesis of ATP (adenosine triphosphate), a high-energy molecule. ATP and reduced NADP that resulted from the light reactions are used for CO_2 fixation in a process that is independent of light. CO_2 fixation involves a number of reactions that is referred to as the Calvin-Benson cycle.

The initial CO_2 fixation reaction involves the enzyme ribulose-1,5-bisphosphate carboxylase/oxygenase (RuBisCO), which can react with either oxygen (leading to a process named photorespiration and not resulting in carbon fixation) or with CO_2. The probability with which RuBisCO reacts with oxygen vs. with CO_2 depends on the relative concentrations of the two compounds at the site of the reaction. In all organisms CO_2 is by far the preferred substrate, but as the CO_2 concentration is very much lower than the oxygen concentration, photorespiration does occur at significant levels. To boost the local CO_2 concentration and to minimize the oxygen tension, some plants (referred to as C_4 plants) have set aside some cells within a leaf (named bundle-sheath cells) to be involved primarily in CO_2 fixation, and others (named mesophyll cells) to specialize in the light reactions: ATP, CO_2 and reduced NADP in mesophyll cells is used for synthesis of 4-carbon organic acids (such as malate), which are transported to bundle sheath cells.

Here the organic acids are converted releasing CO_2 and reduced NADP, which are used for carbon fixation. The resulting 3-carbon acid is returned to the mesophyll cells. The bundle sheath cells generally do not have PS II activity, in order to minimize the local oxygen concentration. However, they retain PS I, presumably to aid in ATP synthesis. Even though C4 plants have reduced amounts of photorespiration, the amount of ATP they need per amount of CO_2 fixed is a little higher than in other plants, and therefore their total production rate is similar to that of plants with higher rates of photorespiration.

Some plants living in desert climates, such as cacti, keep their stomates closed during the day to minimize evaporation (stomates are openings in the leaf surface to enhance gas exchange). These plants take up CO_2 during the night when the stomates are open, and temporarily bind the CO_2 to organic acids in the leaf. During the day the CO_2 is released from the acids and used for photosynthesis. Plants using this mechanism of CO_2 fixation are called CAM (Crassulacean Acid Metabolism) plants.

Increasing CO_2 Levels

The amount of overall CO_2 fixation in plants growing under optimal conditions is limited primarily by the amount of CO_2 available. Therefore, the increase of CO_2 in the atmosphere will lead to somewhat higher rates of plant growth in environments where the CO_2 concentration limits growth rates. This is usually the case in an agricultural setting, where nutrients and water availability are not limiting.

However, also in natural conditions, where limitations other than the CO_2 concentration will generally limit plant productivity, plant productivity has been found to often increase upon increasing the CO_2 concentration.

Photosynthesis and Respiration

Virtually all oxygen in the atmosphere is thought to have been generated through the process of photosynthesis. Obviously, all respiring organisms (including plants) utilize this oxygen and produce CO_2. Thus, photosynthesis and respiration are interlinked, with each process depending on the products of the other. The global amount of photosynthesis is on the order of a trillion kg of dry organic matter produced per day, and respiratory processes convert about the same amount of organic matter to CO_2.

A large part (probably the majority) of photosynthetic productivity occurs in open oceans, mostly by oxygenic prokaryotes. Without photosynthesis, the oxygen in the atmosphere would be depleted within several thousand years. It should be emphasized that plants respire just like any other higher organism, and that during the day this respiration is masked by a higher rate of photosynthesis.

Diversity of Photosynthetic Organism

Even though plants are the most visible representatives of photosynthetic organisms, it should be emphasized that many other types of photosynthetic organisms exist. All photosynthetic bacteria other than the cyanobacteria and their relatives use only one photosystem, and for thermodynamic reasons they cannot use water as the ultimate electron donor.

Instead, they can use reduced compounds such as H_2S as donor. However, CO_2 fixation occurs in these organisms. Some of these photosynthetic bacteria appear to have retained an evolutionary ancient arrangement of their photosynthetic apparatus, and are of interest for the analysis of evolutionary relationships of photosynthetic systems.

An extensive group of these photosynthetic bacteria, the heliobacteria, was discovered rather recently in the 1980s. The first representative of this group was isolated by the group of Dr. Howard Gest from a soil sample collected on the campus of Indiana University, and this isolation was the result of a fortunate coincidence of serendipitous events. Analysis of the heliobacterial reaction center has helped to lay the basis for the current concept that all photosynthetic reaction centers from the large variety of photosynthetic organisms are related to each other.

The majority of bacteria cannot be maintained in pure culture (that is, without other oganisms). This has essentially limited analysis of photosynthetic prokaryotes to the relatively small group of organisms that can be grown in pure culture. It is likely that the actual diversity of photosynthetic organisms is much larger than is known thus far. Indeed, species with novel photosynthetic properties are reported virtually every year.

For example, recently an organism was reported that has chlorophyll d (a chlorophyll that is very rare in nature) as the main pigment. Moreover, several years ago, previously undetected and very small chlorophyll *a/b*–containing prokaryotes were recognized to be the major contributors to photosynthetic production in the open ocean. This emphasizes that much relating to biodiversity and photosynthesis is still to be discovered, and that these discoveries are not limited to tropical rainforests and other ecological settings of large popular interest.

Evolution

In eukaryotes, photosynthesis takes place in the chloroplast, which has long been known to have prokaryotic features. Chloroplasts are thought to have evolved from a cyanobacterium (or close relative) that was in a symbiotic relationship with a eukaryotic, non-photosynthetic cell and was engulfed inside this cell. The cyanobacterium and the eukaryotic cell presumably were in a mutually beneficial relationship (endosymbiosis), with the photosynthetic organism sharing some of its produced carbohydrates with the host, and the host providing the photosynthetic bacterium with other compounds. The prokaryote slowly gave up its independence as well as its cell wall, and some of its genetic information was transferred to the nucleus of its eukaryotic host.

The resulting chloroplast maintains a small, prokaryote-like circular DNA of its own (DNA is material carrying genetic information); this DNA contains the genetic blueprint to make many of the membrane proteins needed in the chloroplast, which apparently are not easily targeted to and/or transported into the chloroplast. Occasionally, photosynthetic organisms are found where the chloroplast has retained a little more of the original cyanobacterial features. For example, in algae such as Cyanophora paradoxa plastids (called cyanelles) are found that resemble cyanobacteria in their overall morphology as well as in the fact that they are surrounded by a cell wall.

Not all chloroplasts have resulted from a single endosymbiotic event, but apparently from multiple events that occurred independently. Chloroplasts from higher plants and many green algae probably all result from the same endosymbiotic event, whereas chloroplasts from red and brown algae and from diatoms are the result of one or more other events. The situation is even more complicated in cryptomonads, a type of algae, and chlorachniophytes, photosynthetic amoebae, which apparently are the result of an endosymbiotic event of a eukaryotic alga in a eukaryotic host. The nucleus of the endosymbiont has been mostly degraded, resulting in a chloroplast enveloped by four membranes.

Early Events

Chlorophyll is used by all photosynthetic organisms as the link between excitation energy transfer and electron transfer. Of particular note is the rate with which these transfer reactions need to occur. As the lifetime of the excited state is only several nanoseconds (1 nanosecond (ns) is 10^{-9} s), after absorption of a quantum, energy transfer and charge separation in the reaction center must have occurred within this time period.

Energy transfer rates between pigments are very rapid, and charge separation in reaction centers occurs in 3–30 picoseconds (1 picosecond (ps) is 10^{-12} s). Subsequent electron transfer steps are significantly slower (200 ps – 20 ms) but, nonetheless, the electron transport chain is sufficiently fast that at least a significant part of the absorbed sunlight can be used for photosynthesis.

However, in the presence of excess light, damage may occur, which may originate from the formation of chlorophyll in "triplet state". In a triplet state two electrons in the outer shell have identical rather than opposite spin orientation. This triplet chlorophyll readily reacts with oxygen, leading to the very reactive singlet oxygen, which can damage proteins. To counter this damaging reaction, carotenoids are usually present in close vicinity to chlorophylls.

Many carotenoids efficiently "quench" triplet states of chlorophyll, thus avoiding formation of singlet oxygen. Chlorophyll in its free form is very toxic in the light in the presence of oxygen, because a close interaction with carotenoids is not always available under such circumstances. Therefore, all chlorophyll in a cell in aerobic organisms is bound to proteins, generally with carotenoids bound to the same protein.

Structure Determinations

Because of the strict requirements of positioning of pigments and electron transfer intermediates to allow efficient electron transfer and minimal damage, the structure of pigment-protein complexes involved in photosynthesis is critical. With the exception of specific antenna complexes (such as phycobilisomes in cyanobacteria and chlorosomes in green bacteria), pigment-binding proteins are usually hydrophobic membrane proteins. This initially hampered attempts to elucidate the structure of these complexes, as membrane proteins do not readily form the well-ordered crystals that are needed for high-resolution X-ray diffraction studies.

However, in the 1980s the first structure of a membrane protein complex, the photosynthetic reaction center from a purple bacterium, was determined at high resolution (about 3 Å; in comparison, the distance between neighboring atoms in a molecule is about 1 Å). Investigators from the Max Planck Institute in Martinsried (Germany) who were involved with this work, most notably Hartmut Michel and Johann Deisenhofer, received a Nobel Prize in chemistry for this research. Since then, the structure of various reaction centers and antenna complexes has been determined at resolutions between 2 and 4 Å.

Similarities Between Reaction Centers

Surprisingly, structural comparison of reaction centers from different photosynthetic systems showed that these reaction centers are basically similar to each other in terms of their overall three-dimensional structure. The basic reaction center unit consists of a protein complex with 10 transmembrane helices originating from either two identical protein subunits or from two similar polypeptides of common evolutionary origin.

Each of these polypeptides contribute five membrane-spanning helices, and bind 2-3 chlorophylls (or, in the case of anoxygenic bacteria, bacteriochlorophylls). The fourth membrane spans from each subunit are held together by two chlorophylls, which are the chlorophylls in the reaction center that can be oxidized (give up an electron) upon excitation. Directly associated with the reaction center are proteins that bind antenna pigments. In the case of PS I and similarly organized reaction centers from green and heliobacteria, the antenna portion, with six transmembrane helices, is attached to the N-terminal end of the reaction center proteins.

Implications of Photosynthesis Studies

Photosynthesis has been studied in significant detail and photosynthetic systems are used frequently for development and application of advanced technologies, because photosynthetic systems are fairly well understood, are complex, and often undergo rather unusual biochemical reactions. Some examples are provided below.

Rapid Electron Transfer Reactions

A major difficulty in measuring enzyme kinetics at relatively short time scales (less than 1 ms) is that "traditional" enzyme reactions require a mixing of substrate and enzyme, which usually takes a relatively long time. Kinetic analysis of light-driven reactions such as photosynthetic electron transport have a great advantage in this respect: reactions can be triggered simply by a light pulse, which can be even shorter than 1 ps. Moreover, many of the components participating in electron transfer have different absorption spectra depending on whether they are in the oxidized or reduced form.

Using laser spectroscopy methods or more standard optical spectroscopy, it is relatively simple to follow the electron around on a timescale between 1 ps and several ms. The primary charge separation occurs in several ps, and reactions become gradually slower as they involve components that are further away from the reaction center. Because of the fast speed of early reactions, the electron and the "electron hole" are physically separated rapidly by a large distance (the electron generally has traveled about 2 nm to the other side of the membrane within 1 ns after charge separation), so that back reactions (charge recombinations) are not favorable anymore.

Unpaired electrons on reactants that are transiently formed during redox reactions involving transfer of a single electron in many instances can be detected using electron paramagnetic resonance (EPR) and derived techniques (including ENDOR, electron nuclear double resonance, and ESEEM, electron spin echo envelope modulation). Many of these techniques can be used to kinetically follow redox reactions, and provide detailed information regarding electron spin distributions etc. Therefore, photosynthetic membranes and reaction centers have a prominent place as experimental systems in biochemistry and biophysics.

Organic Molecules Mimicking Reaction Centers

Reaction centers are essentially an assembly of cofactors, held in the appropriate position and orientation by the protein environment. Several groups have used the natural system as a model to design organic molecules where the equivalents of the different cofactors are linked together by covalent bonds of various lengths. The result is the creation of a number of sophisticated molecules that serve as "artificial reaction centers". The more advanced molecules consist of two chlorophyll-type molecules linked together (one serves as the electron donor, the other as the acceptor), with the electron-accepting molecule linked to two quinones, which serve as electron acceptors in the natural system.

The electron donating chlorophyll analog is linked covalently to a carotenoid, which can donate an electron to the oxidized chlorophyll. Upon excitation of the chlorophyll, a charge separation occurs resulting in an oxidized carotenoid and a reduced quinone. This charge-separated state is formed with high efficiency.

Such molecules can be introduced into liposomes (artificial membrane vesicles) in a specific orientation, and when these are excited by light, a charge separation will occur across the liposome membrane. This results in an electric potential or proton gradient across the liposome membrane, which may be used for a variety of purposes, including ATP synthesis (the latter requires introduction of the ATP synthesizing enzyme into the liposome membrane). The groups of Ana and Tom Moore and Devens Gust at Arizona State University are leaders in developments in this area.

Genetic Modification and Protein Engineering

Because of the ease of detailed functional analysis of reactions and their rates in photosynthetic systems, reaction center complexes are frequently used to determine the consequences of small alterations in the polypeptides on the functional characteristics of the cofactors. Changes at single amino acid residues in the reaction center complex are sufficient to introduce large changes in the properties of cofactors, which in turn leads to altered electron transfer rates and efficiencies.

Single amino acid changes at specific sites in the protein are easily introduced by genetic modification techniques, and resulting functional changes can be studied. An elegant example of such an approach is the modification of the midpoint redox potential of the bacteriochlorophyll in the reaction center of purple bacteria.

The midpoint redox potential is correlated with the ease with which an electron is given off after excitation and is taken up by the oxidized bacteriochlorophyll. Jim Allen, JoAnn Williams and coworkers at Arizona State University found that creating or deleting hydrogen bonds between the protein and the bacterioch lorophyll changed the midpoint redox potential of this bacteriochlorophyll in a rather predictable manner.

In this way, reaction center complexes can be built with different oxidizing strengths, and effects on reaction rates and ultimately the effectiveness of alternate electron donors can be determined. Mutational analysis of photosynthesis proteins is simple in several bacterial systems.

The reasons why this is so in selected cyanobacteria and purple and green bacteria are that:

- Foreign DNA is taken up by the cell spontaneously or is introduced easily by other means such as electroporation ("electric shock"),
- Once the DNA is inside it is incorporated into the organism's genome at one predictable and specific site by means of a process named homologous double recombination,
- The organism can be propagated without relying on photosynthesis, for example using an added carbohydrate source.

Genetic approaches involving directed mutagenesis as described above have proven to be very useful in studying photosynthetic electron transfer and will be of increasing relevance for the design of photosynthetic organisms for biotechnological uses. By this method the function of a large number of genes has been probed, and the role of individual domains and residues has been determined. Genomic sequencing projects are very useful in this respect, and the complete DNA sequence of one photosynthetic organism is already known. From the DNA sequence, the potential of the organism can be determined.

The entire 3,573,470 nucleotide-long genomic DNA sequence of the transformable cyanobacterium *Synechocystis* 6803 was determined by Satoshi Tabata and coworkers at the Kazusa DNA Research Institute in Japan. This organism is used by several researchers, including in the Vermaas group at Arizona State University, to elucidate the role of many proteins thought to be involved in photosynthetic or other physiological processes.

Meanwhile, other groups are working on the genomic sequence of two purple bacteria. With the genomic sequence in hand, the role of specific genes can be found by amplifying the gene of interest by means of polymerase chain reaction, cloning it into a plasmid, replacing the gene by a selectable marker (i.e., a piece of DNA coding for a protein inactivating a particular antibiotic, thus conferring antibiotic resistance), and analyzing the functional characteristics of resulting mutants.

GENETIC MODIFICATION OF HIGHER PLANTS

Globally, improving plant productivity by genetic means has been an important goal for many years. Even though initially it was hoped that crop productivity could be boosted by "improving" the photosynthesis process, it has become increasingly apparent that improved productivity will be best accomplished by genetic modification of crop plants to introduce insect or pathogen resistance, or to yield improved vitality under marginal conditions (for example, at high salinity, which has become a significant factor in many agricultural areas because of continued irrigation).

A pertinent example is the development of varieties of cotton and other crops that express a pro-toxin from a bacterium, Bacillus thuringiensis, that is converted to a toxin in the midgut of particular insects such as caterpillars but not in other life forms. This allows efficient biological control as long as the insects do not develop resistance to this compound.

Genetic modifications to intrinsically and significantly improve the photosynthesis process have not yet been successful. The reason for the apparent inability to "improve" the photosynthesis process itself presumably is related to the fact that photosynthetic systems have evolved over a relatively long period of time, and that the selection factors have not changed significantly in recent history.

This has led to the emergence of a very effective photosynthetic apparatus that is difficult to improve upon by simply changing some amino acid residues or by introducing or deleting some genes. If relatively simple changes could have significantly improved the photosynthesis process per se, Mother Nature would have already found these as natural mutations are rather frequent. However, it is possible that significant progress in this area can made in the future, if new design paradigms for enzyme function and specificity can be developed.

For example, if protein structures (particularly the structure of the active site) can be better modeled and predicted, one should be able to further improve upon the RuBisCO specificity of CO_2 vs. oxygen. It is also important in this respect to determine what the rate-limiting step in the process is under natural conditions. Even though more effective light capture by crop plants might be considered by introduction of antenna pigments that absorb in the green and yellow region of the light spectrum, light capture and the light reactions usually are not limiting plant productivity in agricultural settings. Therefore, such modifications in a plant will result in increased productivity only in light-limiting settings.

Future Directions

Light energy is cheap, clean, and essentially inexhaustible. With limited supplies of fossil fuel and increasing concern about CO_2 emissions, further development of technologies that make use of solar energy is inevitable. Current silicon-based technologies for the harvesting of solar energy require a very energy-intensive production process and even though they have improved significantly over the years in their efficiency, further development of photosynthesis-based technologies for energy collection is certainly warranted.

However, photosynthesis and related processes can be applied to many more areas than just solar energy conversion. Realizing that novel designs and applications of light-mediated processes have enormous promise in the next decade and beyond, Arizona State University has started an initiative, Project Ingenhousz, named after the 18th century physician who discovered that light is needed for oxygen evolution by plants, and that only green parts of the plant carry out this process. The goal of the initiative is to capitalize on research progress and ideas in this area, and to more effectively interface academia, where many of the discoveries are made, with the private sector, where such discoveries are worked out further and applied.

For example, there are a myriad of possible applications of artificial reaction centers and related molecules in nanotechnology. Many synthetic pigments also have found biomedical uses in tumor detection, as they -for unknown reasons- tend to accumulate preferentially in tumors and are highly fluorescent and thus easily detectable in a patient whom is being operated on to surgically remove a tumor.

In the biotechnology field, photosynthetic organisms are likely to play an increasing role in (over)production of enzymes, pharmaceuticals, nutraceuticals, etc., which until now are produced primarily by genetically modified heterotrophic microorganisms such as yeast and selected bacteria.

A major advantage of photosynthetic organisms is that no fixed-carbon source needs to be added for growth and, therefore, production costs are lower and the chances of contamination with other microorganisms are less. There are several ways to modify organisms to have them (over)produce useful compounds. One is to introduce specific genes under a strong promoter, leading to high expression of these genes and to synthesis of a "new" enzyme. Another way is to delete genes so that substrates will accumulate. For this to be optimally successful the metabolic pathways of an organism need to be fairly well understood, and genomic DNA sequences are an important step in this direction.

A third way to produce a new compound is to utilize an existing enzyme and to modify the site where the substance to be converted (the substrate) binds, so that a different substrate can be bound and a different product can be formed. With any of these approaches, selection for randomly generated combinatorial mutants with specific properties also has been proven to be effective. With an increasing arsenal of genomic sequences, and with improving knowledge regarding determinants of protein structure and cofactor binding, any of these three approaches are very promising.

Another potential application of photosynthetic organisms is in bioremediation. Bioremediation is the clean-up of environmental (soil or water) pollutants by biological means. An example is the biological breakdown of toxic organic compounds into innocuous products. Also, remediation of nitrate from drinking water supplies is becoming an increasingly pressing issue. The advantage of using photosynthetic organisms is that no external energy source needs to be provided for growth of the organism if it is in the light, making these organisms very suitable for remediation of aqueous surface environments.

Another utilization of photosynthetic organisms is to have these organisms use solar energy to produce clean-burning fuels. Even under natural conditions some photosyn thetic systems such as algae can produce hydrogen, which probably is the cleanest fuel as it reacts with oxygen to produce water.

However, the cheapest and most universal electron donor of all, water, upon oxidation in PS II forms oxygen, which is not very safe in combination with hydrogen. For this reason methods have been sought to temporally separate oxygen evolution and hydrogen production in algae; in the laboratory this can be achieved by sulfur deprivation, which preferentially inhibits photosynthesis. Another option would be to use photosynthetic organisms for methane production. Even though methane upon combustion will form CO_2, the overall atmospheric CO_2 balance would not be disturbed as an equal amount of CO_2 will have been taken out of the atmosphere upon methane production by the photosynthetic organism.

Research in photosynthesis in all its facets has proven to have opened many doors in a variety of disciplines, ranging from biophysics to plant physiology. Progress has been driven by an interdisciplinary approach to this complex, yet fascinating, spectrum of problems, challenges, and opp ortunities. Photosynthesis is the basis of our food and energy supply, and innovative utilization of solar energy is likely to be of increasingly critical importance in the future. This, together with novel uses of photosynthetic principles for other purposes, make it likely that photosynthesis and its appli cations will help to shape an increasingly broad area of exciting discoveries and innovative ideas.

12

Factors Influencing the Wilting Coefficient

SOIL TYPES AND TESTING

Soil can be described in many different ways, such as heavy, light, sandy, clay, loam, poor or good.

Scientists typically describe soil according to its:

- Colour
- Compaction
- Moisture content
- Organic content
- pH
- Profile
- Structure
- Temperature, and
- Texture

Although each of these factors is important, three factors (texture, organic content and pH) are more important than the others.

COLOUR

Soil colour can provide information about organic matter in the soil, drainage, biotic activity, and fertility. To identify the colour of your soil, you should take a garden spade or shovel, and dig a shallow hole, at least 3" - 4" deep, and gauge the colour (you should do this quickly before the sun can dry it out).

	Colour		
Condition	Dark	Moderately dark	Light
Organic matter	High	Medium	Low
Erosion factor	Low	Medium	High
Aeration	High	Medium	Low
Available nitrogen	High	Medium	Low
Fertility	High	Medium	Low

COMPACTION

To be healthy, a soil needs to be able to breath and water needs to be able to move through it reasonably easily. Compacted soils don't allow much air to circulate to the root zone and water (rainfall or irrigation) tends to just run-off. This increases erosion and strips away vegetation and topsoil. A normal, loosely compacted soil helps to absorb and retain water, releasing it slowly, and allows the root zone of plants to "breath". These soils are generally more productive, since plants can grow much more readily. Dense, highly compacted soils typically have less plant growth, which increases run-off.

The rate of infiltration of water is an excellent indication of soil health.

You can measure the water infiltration rate very easily:

- First, get a large, empty coffee can and cut off the bottom.
- Second, beginning about 3" up from the bottom, mark the inside of the can every ?" with a permanent marker, being careful not to cut your hand on the edges of the can.
- Drive the can about 3" into the ground until the first mark is level with the ground (placing a board on the top of the can and pounding the board with a hammer will help drive the can into the ground. Be careful not to irrigate the area first, since this will prevent you from getting an accurate measurement of the infiltration rate.)
- Fill the can with water clear to the top and begin timing the rate of infiltration. Measure the amount of water that has drained into the soil at the end of each minute for the first ten minutes. Determine the rate of infiltration in inches per minute by dividing the total number of inches of water that drained away in the can by 10 minutes. Knowing the actual water infiltration rate for your yard is critical if you want minimize the amount of water you use.
- Repeat the experiment at several areas around your yard, being careful to record each location and its infiltration rate. If the infiltration rates at each location vary considerably, then draw a quick sketch of your yard, and plot the infiltration rate for that area. If you install an automated sprinkler system, you can adjust the emitters in each area to only deliver the amount of water that can infiltrate in a given amount of time. This will eliminate irrigation run-off from your yard or garden, while ensuring adequate soil moisture for plants.

Sand is the largest particle in the soil. When you rub it, it feels rough. This is because it has sharp edges. Sand doesn't hold many nutrients or water. Silt is a soil particle whose size is between sand and clay. Silt feels smooth and powdery. When wet it feels smooth but not sticky. Clay is the smallest soil particle. Clay is smooth when dry and sticky, or plastic when wet. Soils high in clay content are called heavy soils. Clay can hold a lot of nutrients, and some kinds can hold quite a bit of water, but the structure of clay doesn't let air and water move through it well. Most of the water in a clay soil is so tightly bound to the clay particles that plants can't get it loose.

Moisture

The amount if moisture found in soil varies greatly with the type of soil, climate and the amount of humus (organic material) in that soil. The types of organisms that can survive in your soil is largely determined by the amount of water available to them, since water acts as a means of nutrient transport and is necessary for cell survival. Soil moisture can be estimated visually, although this is quite imprecise. Soil moisture can also be determined by a soils laboratory. Soils labs typically dry a sample in an oven or on a hot plate (approximately 225? F for 24 hours) and compare the weight of the soil before drying to the weight after drying. The moisture content is reported as per cent moisture on a weight basis. Several irrigation system manufacturers have developed soil moisture indicators that can be used to control irrigation more precisely, turning the system on only in areas where more water is needed and then only for the minimum time necessary to get the soil moisture back up to the desired level.

Organic Content

The organic content of soil greatly influences the plant, animal and microorganism populations in that soil. Decomposing organic material provides many necessary nutrients to soil inhabitants. Without fresh additions of organic matter from time to time, the soil becomes deficient in some nutrients and soil populations decrease. The amount of organic material can be determined by ignition. Organic material is made of carbon compounds, which when heated to high temperatures are converted to carbon dioxide and water. In the ignition process, a dry solid sample is heated to a high temperature. The organic matter in the soil is given off as gases. This results in a change in weight which allows for calculation of the organic content of the sample.

Oven-dry the sample to remove water. Weigh a crucible and lid, evaporating dish and cover, or other covered container. Place approximately 10 grams of soil sample in the container, cover it and weigh the sample, container and cover. Place the container on a metal stand and heat it with a propane torch. Allow the fumes to escape, but not the soil particles. Heat the sample strongly after most of the gases have escaped; continue heating until there are no visible fumes. Cool the container, lid, and sample. Reweigh and calculate the per cent of organic material.

Soil pH

Most people think that rainwater has a pH of 7, so it comes as something of a shock when they learn that rainwater (if its not polluted) has a normal pH of about 6 - 6.5, which is slightly acidic. This is due to dissolved carbon dioxide from the air, which reacts with water to form a dilute acid (carbonic acid), much like the carbon dioxide in soda. It should then come as no surprise that most plants grow their best at around the same pH*. You can determine the pH of your soil very easily using a universal indicator solution or pH paper, available at most hardware stores in the pool supplies section. To determine the pH, just put a small amount of the soil to be tested in a clear or white container, being careful not to touch the sample. Pour a small amount of universal indicator over the soil, then match the colour *of the indicator solution* (not the soil) with use pH paper, pour a small amount of water on the soil sample.

Touch the pH paper to the sample and match to colour of the paper to the pH colour chart. Some plants, such as rhododendron, camellias, azaleas, blueberries, ferns, spruce, pines, firs, and red cedar prefer soil that is more acidic, with a pH of 4.0 to 5.0. Other plants, such as beech, mock orange, asparagus and sagebrush tolerate soils with a pH 7.0 to 8.0. Above a pH 8.5, the soil is too alkaline for most plants, while if the soil pH is below 3.5 it will be too acid. You should also note that each layer of soil may have a different pH, which means that pH can vary within the soil, although the differences are usually not too great.

Soil Profile

If you really want to know about your soil, the best way to start is to obtain a cross-section of the various layers. This can be done fairly easily if you use a soil core tool. A soil core tool is little more than a hollow tube 2 to 4 feet in length with a handle and cross piece like a shovel to help push it in. Once the tool has been inserted into the soil, it should be turned to loosen the soil and then pulled out. The resulting soil core can be easily examined to identify the various layers (each layer is also called a horizon) in the soil, the aggregate of which is called a soil profile. To determine a soil horizon, you simply mark where the soil changes colour and/or general appearance.

Many soils have three major layers or horizons, top soil, subsoil and parent material. Depending on where you sample, the top zone may be comprised of actively growing plants and dead plant materials. The top soil is typically darker coloured and usually has more organic matter, higher biotic activity, abundant roots, and commonly lower in nutrients than underlying layers. The first inch of top soil may be lighter in colour because many of the nutrients may have been leached out by water, and organic material may have been partially oxidized by sunlight and heat. The soil immediately below the first inch is usually somewhat darker, has many roots, moderate organic matter, and provides most of the nutrients for the plants. The next major layer is the subsoil. This layer is typically 1 to 2 feet below the surface and is characterized by a lighter colour with much fewer, larger roots. The subsurface layer generally has less clay than the topsoil. The third layer, which may not be observable, is the parent material. This consists of unconsolidated, slightly weathered rocky materials from which soil develops. It is characterized by limited biotic activity and very few roots.

Soil Structure

Soil structure tells how the soil affects the movement of water, air and root penetration into the soil. The geometric shapes of the soil determine how it is put together. Words such as blocky (the blocks of soil are large, with the same number of cracks going horizontal as vertical), granular (the blocks of soil are small, with the same number of cracks going horizontal as vertical), columns (the blocks of soil and related cracks are generally longer in the vertical direction than in the horizontal), and plate-like (the blocks of soil and related cracks are generally longer in the horizontal direction than in the vertical), describe soil structures. To determine the structure of your soil, carefully break apart each layer and match its characteristics with the appropriate structural type.

Structural type	Water penetration	Drainage	Aeration
Columns	Good	Good	Good
Blocky	Good	Moderate	Moderate
Granular	Good	Best	Best
Plate-like	Moderate	Moderate	Moderate

Soil Temperature

Soil temperature has a significant role in helping to determine the rate of plant growth, and whether a plant will even survive. The temperature in your soil changes greatly with depth. To measure soil temperature, find an area that is not in direct sunlight. Using a thermometer, measure the air temperature at shoulder height. Hold the thermometer still for about one minute (make sure your fingers are not on the thermometer bulb), read and record the air temperature. Next, measure the temperature at the surface of the ground.

Put the thermometer flat on the ground and record the temperature after one minute. To determine the temperature below the ground surface, use a dowel that you have marked at 1 inch, 2 inches, 6 inches and 12 inches. Start by pushing the dowel into the ground till you reach the 1 inch mark. Remove the dowel and insert the thermometer for one minute, then remove the thermometer and quickly record the temperature. Repeat this procedure to obtain temperature readings at 2 inches, 6 inches and 12 inches. Take temperature readings at different times throughout the day at the same location. To compare with soil temperatures for areas in direct sun, just repeat using the same procedure but select an area that gets full sun. You will note that the soil temperatures in these areas are typically much higher than in the shaded areas.

Soil Temperature	Conditions during growing season
Less than 40°F	No growth, bacteria and fungi Arenot very active
40°F to 65°F	Some growth
65°F to 70°F	Fastest growth

70°F to 85°F	Some growth
above 85°F	No growth

Soil Texture

Sandy soil absorbs more than two inches of water per hour. It is very porous, with large spaces between soil particles. Little water is retained and the sandy soil dries out quickly. Loam soil absorbs from.25 inches to 2 inches per hour.

The soil is loose and porous and holds water quite well. Clay soil absorbs less than.25 inches of water per hour. Clay soil is dense with few air spaces between particles and holds water so tightly that little water is available for plants.

Characteristics of Different Soil Types

It can be argued that no two soils are ever exactly alike. Although this is true, it is useful to group soils into categories. Three major categories of soil dominate our area.

These are:

- Sandy soil
- Loam soil, and
- Clay soil

To figure out what type of soil you have, there are several easy methods. The first, called the rope test, requires that you squeeze a moist, but not muddy, one inch ball of soil in your hand. Then rub the soil between your fingers. Sandy soil feels gritty and loose. It won't form a ball and falls apart when rubbed between your fingers.

Loam soil is smooth, slick, partially gritty and sticky and forms a ball that crumbles easily. It is a combination of sand and clay particles. Clay soil is smooth, sticky and somewhat plastic feeling. It forms ribbons when pressed between fingers. Clay soil requires more pressure to form a ball than loam soil, but does not crumble apart as easily.

A second test is called a jar test and is very easy to do. Here's what you'll need:

- 1 clean quart jar and tight fitting lid
- Clean water
- Soil sample

First, find an empty, clean quart jar (an old mayonnaise jar works very well for this test.) Fill the jar about 2/3 full with clean water.

Next, take a sample of soil (break the large clods apart so it will fit through the jar opening) and fill the jar and water until the jar is nearly full, leaving about?" of air space at the top. Screw on the lid and shake it vigourously for a minute or two, until all the soil particles are broken down into suspension in the water.

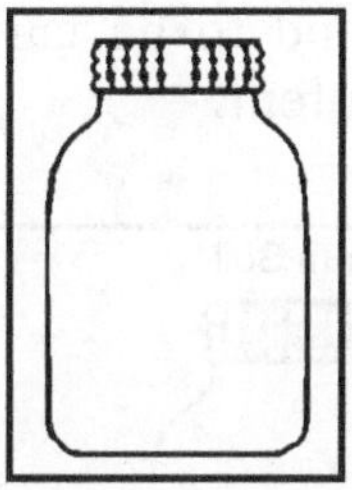

Now, allow the suspended soil to settle for about a minute, and place a mark on the side of the jar at the top of the layer that has settled out.

This is the sand layer is comprised primarily of sand and larger particles. Set the jar aside, being careful not to mix the sand layer that has already settled and wait approximately an hour.

Now, place a mark on the side of the jar at the top of the next layer to settle out. This is the silt layer.

Again, place the jar aside for a full day, being careful not to shake or mix the layers that have settled out. After 24 hours, or when the water is once again clear (more or less), place a mark on the side of the jar at the top of the final layer. This is the clay layer. The percentage of each layer tells you what kind of soil you have.

- *Type of Soil:* Sandy soils are found throughout Southern California, but are very common near the mountain foothills, along rivers and streams and certain coastal areas. Sandy soils are typically comprised of approximately 80 - 100per cent sand, 0 - 10per cent silt and 0 - 10per cent clay by volume. Sandy soils are light and typically very free draining, usually holding water very poorly due to very low organic content.
- *Example of Test Jar*:

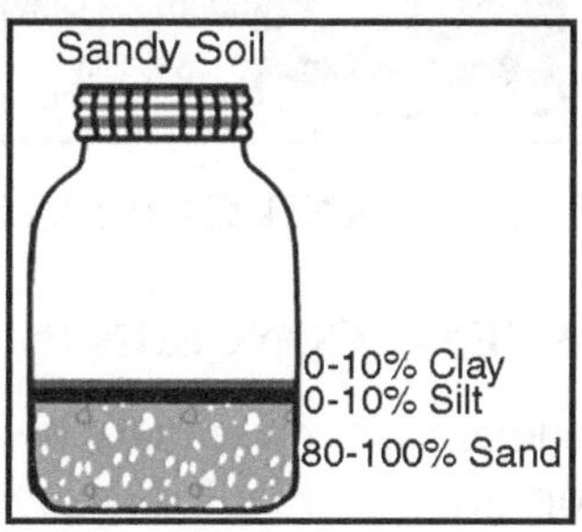

- *Type of Soil:* Loam soils are also common in Southern California, particularly in the valleys and flat areas (flood plains) surrounding rivers and streams. Loam soils are typically comprised of approximately 25 - 50per cent sand, 30 - 50per cent silt and 10 - 30per cent clay by volume. Loam soils are somewhat heavier than sandy soils, but also tend to be fairly free draining, again, due to typically low organic content.
- *Example of Test Jar*:

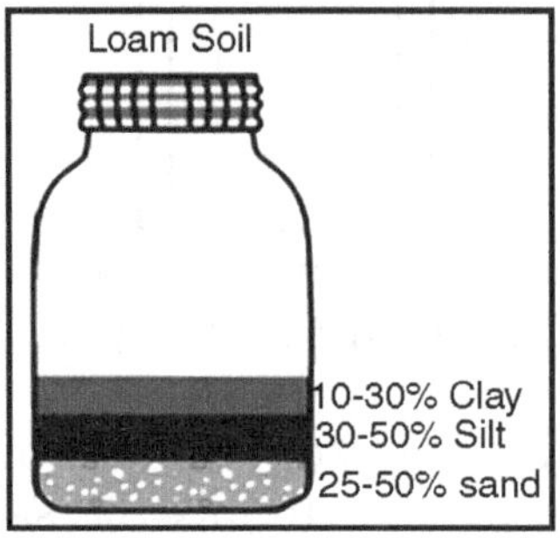

- *Type of Soil:* Clay soils are very common in certain areas, particularly around urban areas where fill soils have been used to establish grade in subdivisions and developments. Clay soils are typically comprised of approximately 0 - 45per cent sand, 0 - 45per cent silt and 50 - 100per cent clay by volume. Clay soils are not typically free draining, and water tends to take a long time to infiltrate. When wet, such soils tend to allow virtually all water to run-off. Clay soils tend to be heavy and difficult to work when dry.
- *Example of Test Jar*:

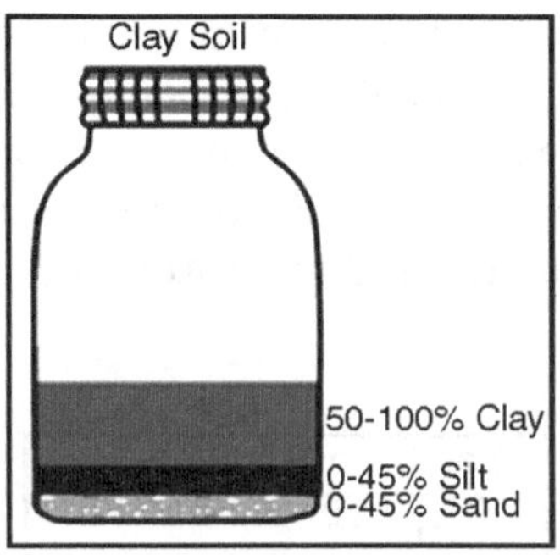

PLANTS AND CLIMATE

WHY IS CLIMATE CHANGE OF CONCERN IN AGRICULTURE?

Agricultural success is dependent on climate because heat, light, and water are the main drivers of crop growth. Plant diseases and pest infestations, as well as the water needed for irrigation, are also governed by climate. The total amount of food produced in the world varies from year to year, largely as a result of weather conditions. Some agricultural regions are more sensitive to changes in weather than others. The most sensitive regions are often in the developing countries where less advanced technology is available to prevent floods and to minimise the impact of droughts and where soils, terrain and climate are less suited to agriculture.

WHAT IS THE MAIN TASK NOWADAYS?

The main task today is to identify the regions which are most likely to be vulnerable to climate change. This will, hopefully, allow regions to adapt so that the impact of climate change is minimised.

PLANTS AND THE ENVIRONMENT

As plants interact with the environment directly by exchanging water and energy, they are very sensitive to storms, droughts and floods. These weather events can severely damage crop yield. In this unit we look at how plants are affected by changes in temperature, humidity and rainfall. Probably the most important role of plants in the environment is the production of oxygen (O_2) and the absorption of carbon dioxide (CO_2) from the atmosphere during the process of photosynthesis.

Photosynthesis is the basic process for plant life. Plants also affect and change their surroundings to make them more suitable for growth. Some of the more minor ways in which plants interact with their environment.

Environmental conditions, such as light intensity, temperature, water availability and wind strength, affect plant growth. Plants also modify the environment around them, they release water which cools the air, breakdown the soil to make it suitable for their roots and for other plants and animals and decrease the speed of the wind. Plants need water to live and grow. High temperatures reduce the availability of water and decrease crop yields. Young plants are especially vulnerable to extreme weather conditions. Let´s now look at some of the mechanisms plants use to cope with extreme weather events such as high temperatures, floods or droughts.

HIGH TEMPERATURES

High temperatures affect crops directly by increasing the rate at which they loose water (their evaporation rate), in the same way that high temperatures make us sweat. Plants have very small pores, called stomata, spread over their leaves. These stomata help the plants control how much water they contain. The stomata are made up of two guard cells which open or close the pore depending on how much water the plant needs. During dry periods, the guard cells are closed so that the plants do not loose too much water. Under normal weather conditions, the stomata are open.

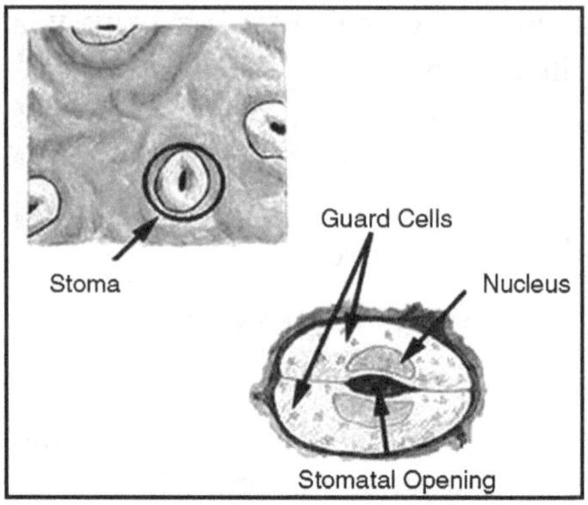

Fig. Function of stomata

Each plant type has different structural characteristics and so all plants don't grow well at the same temperature.

When the optimal temperature value for a particular plant is exceeded, the plant tends not to grow as well leading to a drop in yield. Most plants are very sensitive to high temperatures, although the extent varies depending on the age of the plant and its ability to withstand poor situations. High temperatures also increase the amount of water lost from the soil by evaporation. This also affects plant growth as the soil is their main source of water.

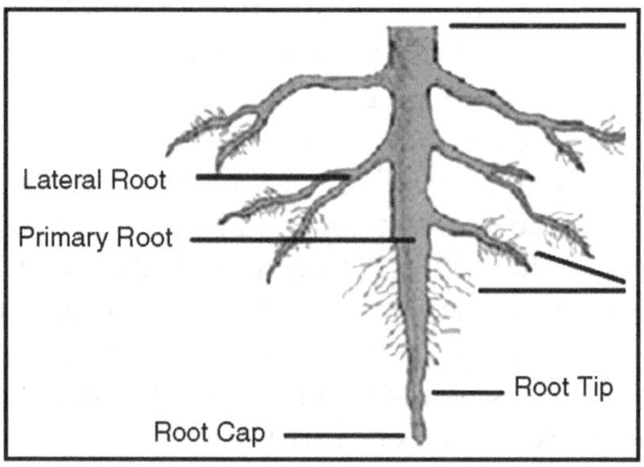

Fig. Structure of Plant Roots.

PRECIPITATION

Precipitation (rainfall) is the primary source of soil moisture, and rainfall amount is probably the most important factor determining the crop yield. A change in climate can cause an increase or a decrease in the amount of precipitation which falls. Roots are one of the main ways plants get water from the environment.

In many parts of the world, plants have much longer roots than trunks or branches.Sometimes a bush that is only around 30 cm high can have roots that go as deep as two meters into the soil. This happens in places where there is not much rain during the year, like the deserts or the very arid dry regions of the world. Plant roots spread throughout the soil so that water can be taken up in different places at the same time. The main roots have fine root hairs on them to allow water to be absorbed easily.

Dry periods can badly affect plant growth but the amount of damage depends on the ability of the plant to expand its root system and how much water the soil can hold onto.

High humidity, frost and hail may also damage certain crops. High temperatures are normally accompanied by dry periods and both the warmth and the lack of water can have negative effects on plant growth. Roots are unable to find water in the soil and stomata have to close to prevent water loss from the plant.This causes the temperature of the plant to rise and this can damage the plant. When plants struggle to grow because of high temperatures or lack of water, they are said to be under stress.Excessively wet years, on the other hand, may cause crop yields to fall.

The soil becomes water logged and the plant roots rot in the excess water. Intense bursts of rainfall may also damage young plants, both because of the hard impact of the rain drops on the tender plants and because heavy rain can cause soil erosion.

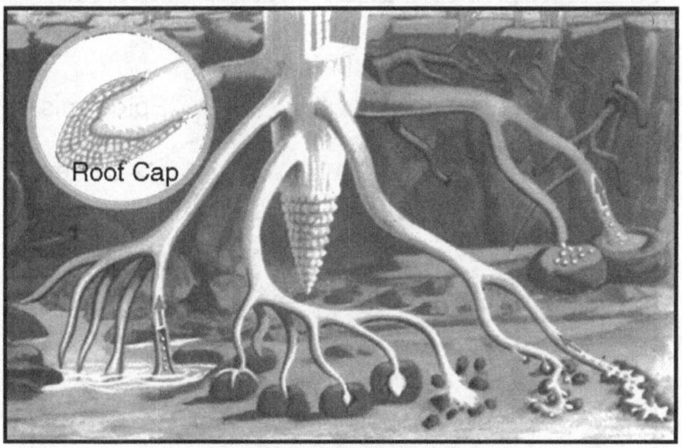

Fig. Corn in the Rain

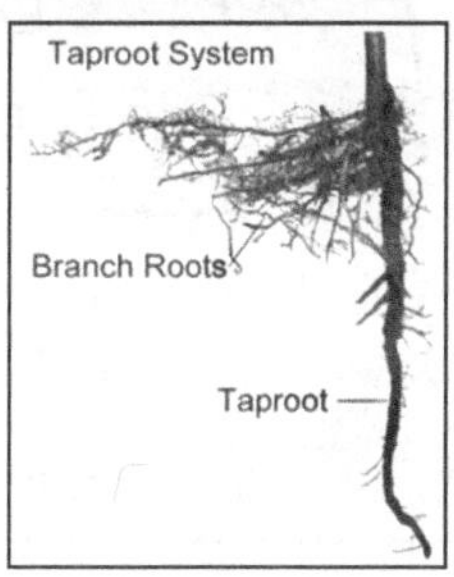

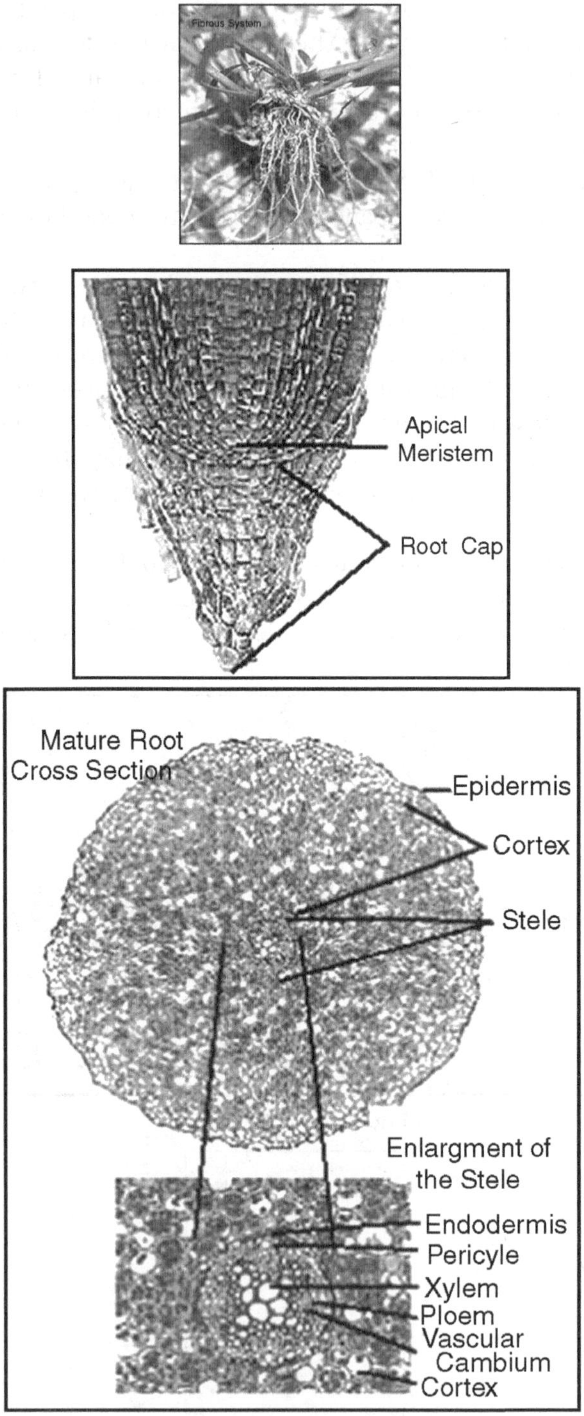

Fig. Plant Root in Action.

Rootlets bury into the earth, absorbing, by osmosis, water and the mineral substances dissolved in it. Roots play an important role in this way of life. They serve as organs of attachment, anchoring the plant to the ground, and also as organs of absorption and transport for water and dissolved salts. The tip of the root, called the root cap, has the task of forcing its way into solid earth. For this purpose the root cap is made up of layers of flattened and hardened cells. Behind the root cap is a thick white down consisting of thousands of tiny hairs. This part of the root has the function of absorbing water, with mineral salts dissolved in it, from the soil. The water is absorbed by a process called osmosis. The liquid inside the root hairs is a rather strong solution of sugars and mineral salts; that outside them, *i.e.*, the water in the soil, is a weak solution of mineral salts.

The outer skin of the root hairs is a membrane, of a type known as a semipermeable membrane, with the special property of enabling weaker solutions pass through into stronger ones. Therefore, so long as the solution inside the root hairs is stronger than that outside, they will take in water. As well as taking in water and salts, roots sometimes give out substances in solution. For instance, if the root meets a hard substance such as marble, which is insoluble in water, it gives out carbon dioxide. When mixed with water, this carbon dioxide acts on the calcium carbonate in the marble and turns it into soluble sodium bicarbonate, which the root can then absorb.

DESCRIPTION OF ROOT

ROOT GROWTH

Normally, roots grow downward under the influence of gravity and away from light. This reaction enables a sprouting seed to get a hold on the soil.

There are two basic patterns of root growth:

1. The tap-root system found, in dandelions
2. The fibrous system found, in grasses.

The tap-root system is an extension of the primary root (radicle) of the young seedling, but in fibrous systems this primary root is quickly displaced by numerous fine roots from the base of the stem. All roots which do not grow from the primary root (*e.g.*, the roots on a rhizome) are called adventitious. Propagation of plants by cuttings relies on the development of adventitious roots on the stem.

ROOT STRUCTURE

At the tip of all roots is a mass of cells–the root cap–which protects the growing point during its passage through the soil. The cells of the root cap are produced by the actively dividing cells of the growing point. as the outer parts of the root-cap wear away, fresh cells replace them. The old cap cells may also lubricate the tip as it grows downward. Cells formed just behind the tip lengthen rapidly and push the tip further into the soil with considerable force. The growing region is followed by the root hair region. The hairs are outgrowths of the epidermis and are the main organs of absorption. They penetrate the soil and absorb water. The root hairs occupy only a limited region. Hairs in the growing region would be sheared off as the root pushed downward. Each hair has only a brief existence and as the hairs die off they are replaced with new ones further down. The hair region is thus kept at a more or less constant size. Branches–if any–occur behind the root-hair region. The root contains the same sort of tissues as the stem but the strengthening tissues of the root are centrally placed, reflecting the pulling strain suffered by the root as opposed to the bending strain imposed on the stem. The inner layer of the wide cortex is called the endodermis. The walls of this layer become thickened with a corky substance. Within the endodermis is the stele whose outer layer is of parenchyma cells and is called the pericycle.

The protoxylem is on the outside of the xylem tissue (exarch condition) which is star-shaped as a rule. Monocotyledons (*e.g.*, grasses) usually have more "arms" than dicotyledons (*e.g.*, buttercups). A central pith occurs in some species. Secondary growth (with very few exceptions) occurs only in dicotyledons. A strip of parenchyma inside the phloem becomes active, forming a cambium which grows and makes contact with the pericycle. A continuous wavy ring of cambium is thus formed. The secondary tissues are produced from this. Opposite the xylem groups a ray of parenchyma is produced instead of vascular tissue. If secondary growth is excessive (*e.g.* in trees) the whole pericycle becomes active and produces a layer of cork outside the stele. The cortex then dies away leaving this corky covering.

BRANCHING IN THE ROOT

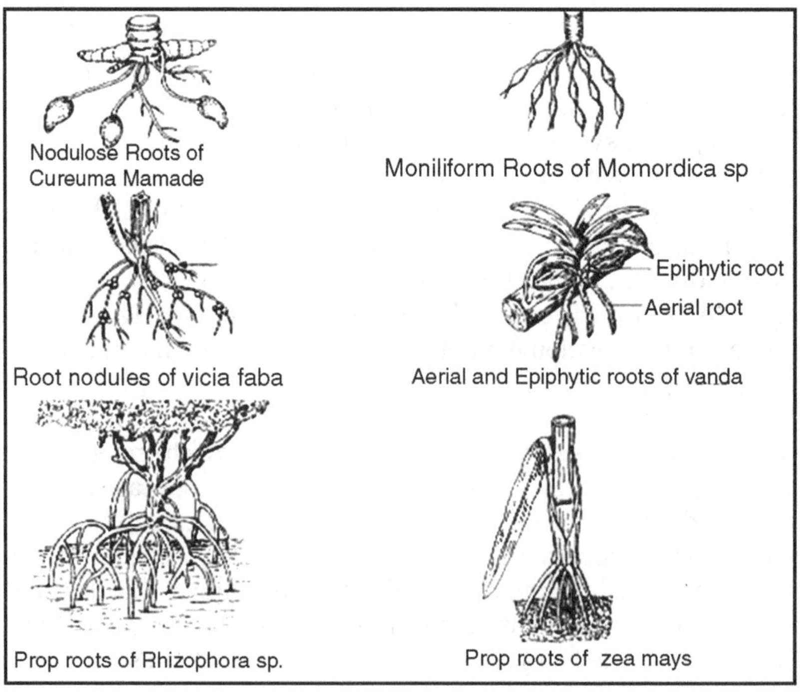

In the stem, buds develop from the outer tissues close to the tip. But if this happened in the root the buds would be torn off as the main root moved through the soil. Root branches develop behind the growing region and are endogenous (*i.e.*, they develop internally). Branching develops before secondary thickening begins. Cells of the pericycle opposite the protoxylem groups become active and produce a growing point just like that of the main root. This new tip grows through the cortex just as the main root grows through the soil. When it breaks out of the parent root the vascular connections are complete and the region is behind that of the root-hairs.

The degree of branching is associated with the size and habit of the plant. Large trees have thick, spreading roots for firm anchorage. Between the root and the stem is a region called the hypocotyl. In this region the vascular tissues change from the root arrangement to that of the stem, but they are continuous all the way through.

ROOT MODIFICAMTIONS

Very often the root is modified for storage. Roots which survive the winter contain food material (*e.g.*, starch) which will be used by the developing shoots in spring. Man has made use of such stores in plants like carrots, turnips, and beet. These are tap roots, but adventitious roots also serve as storage organs. Dahlia tubers and various orchid roots are examples. The food may be stored in the cortex or the phloem.

The climbing roots of ivy are adventitious–arising all the way along the stem– as also are the roots on strawberry runners. Some tropical orchids which grow on tree-trunks have spongy roots exposed to the air. The roots absorb moisture and may contain chlorophyll too. Maize plants and many others have stilt roots which develop from nodes on the stem and provide extra support. Prop roots are used by the banyan tree. Roots develop in the horizontal branches and grow into the soil. They form solid supports for the spreading branches. Some swamp-growing plants such as mangroves develop breathing-roots. The swampy soil is poor in oxygen and some root branches, which may or may not be adventitious, grow upwards into the air. Oxygen diffuses into these and into the rest of the root system.

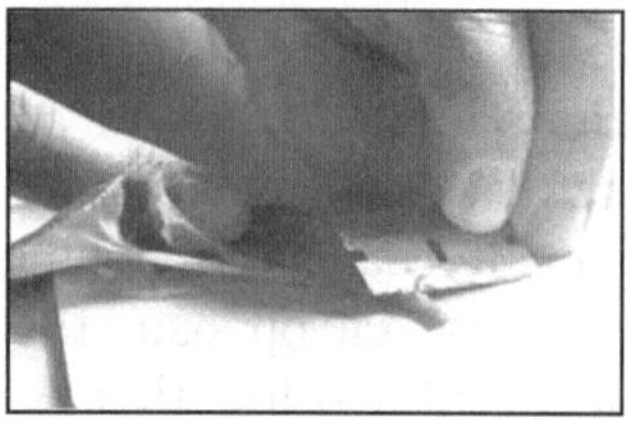

Fig. Root Cutting

ROOT PLANT CUTTINGS

Rooting materials: Clean, coarse, construction-grade sand is suitable for rooting many cuttings. It is also excellent mixed with an equal volume of peat moss. Vermiculite is a lightweight material used for rooting. It holds water well and promotes fine root growth. Perlite is another excellent propagation material. It is lightweight and provides good aeration for rooting. Perlite makes one of the best rooting materials when mixed with an equal volume of peat moss. Don't use field soil as a rooting medium. It packs too tightly under wet conditions and is prone to develop diseases.

Compressed peat pellets are available for seeding and can also be used for rooting cuttings. They expand rapidly when soaked in water. Place them in plastic bags after soaking and draining; insert a single cutting in each pellet and close the bag at the top. No additional watering is necessary until the cutting is rooted and the bag opened.

Pots, medium and equipment used for rooting cuttings must be clean and sterile. Pots should be washed thoroughly using a household cleaner and disinfectant. Tools also should be washed in the same solution or dipped in alcohol. Any rooting medium which is not known to be sterile can be moistened and heated thoroughly in an oven at 150 to 200 degrees for 20 to 30 minutes. Normally, peat moss, vermiculite and perlite don't need sterilization when new.

INSERTING THE CUTTING

Promptly place the prepared cutting in the rooting material; stick the base of the cutting 1 or 2 inches deep, depending on the length of the cutting. Firm the material around the base and settle the medium by watering.

CARE OF CUTTINGS

Never allow the propagation medium to dry out during the rooting process. Since the cuttings have no root system, a high humidity must be maintained around them at all times. Clear plastic is inexpensive and easy to use for covering the cuttings. A plastic bag slipped over a pot is simple and airtight. Support the plastic with wire loops or stakes if need be to keep it from resting on the leaves.Never place a plastic-covered container in direct sunlight. Too much heat will build up under the plastic and burn the foliage.

CARE OF ROOTED CUTTINGS

The length of time needed for cuttings to form roots differs greatly among plants. Check the cuttings occasionally by carefully removing a few from the medium. When a cutting has roots at least an inch long, transplant it into a separate container. The move from high humidity and moist rooting conditions to lower humidity and drier soil is the most critical step in successfully growing new plants from cuttings.

Give these young plants close attention the first few weeks after the move. A good potting medium designed for houseplants can be found at local garden centers or mass erchandisers and will be suitable for potting newly rooted cuttings. After a cutting has become established in the medium, apply a soluble houseplant fertilizer according to directions. Then fertilize at monthly intervals. When the cutting is growing vigourously, normally in spring and summer, fertilizer may be applied every two weeks.

DIVISION

Division is the easiest method of multiplying plants that naturally produce offsets or basal shoots. These new shoots usually have a few roots and can be separated and planted individually. Some plants suitable for division are.

LAYERING

Layering is a method of rooting a new plant while the stem is still attached to the parent plant. It is simple to perform and can be done in the home without special equipment or structures. Methods of layering include simple, tip, air and compound. Air layering is the most suitable method for use on houseplants. It is convenient to do in the home and can be used with plants that would be difficult to root by any other method. Mature wood, about one year old, is generally best for air layering. Old branches or immature wood often root poorly or not at all. The air layer may be made at any point on a stem of proper maturity. On many plants a convenient location is 12 inches from the tip.

AIR LAYERING

Remove all leaves several inches on either side of the point where the layer is to be made. From the center of the layering area, make a slanting cut upward an inch or more in length and about halfway through the branch. A preferred method of wounding is removing a strip of bark 1/2 to 1 inch wide around the branch.

Apply a rooting hormone to the wounded surface or cut. If a cut has been made, don't let it heal back together. Insert a small piece of wood such as a toothpick in the cut to keep the wound open. Take a handful of unmilled sphagnum moss that has been soaked in water and squeeze out excess water. Pack the moist sphagnum around the branch to cover the wound. Cover the ball of moist moss with plastic wrap. An 8- by 10-inch sheet is generally large enough. Wrap it around the moss so that it overlaps and will not allow the moss to dry out. Clear plastic permits you to see when roots have developed. Secure the plastic at each end with electrical tape, string, plant ties or other convenient fasteners. It will usually take a month or more before roots appear.

COMPOUJND LAYERING

Compound layering is suitable for long vines that may be alternately covered and exposed. Wounds should be made on the lower portion of each curve. After rooting, the branch can be cut into segments, each containing its own roots.

CARE AFTER ROOTING

Root systems of newly rooted layers are small in relation to the canopy. After they are severed from the parent plant and potted, the humidity must be kept high. Enclose them in a loose, clear plastic bag for the first week or until they are well established and do not wilt excessively.

GROW MORE PLANTS WITH ROOT CUTTINGS

The major benefit of root cuttings is the simplicity of the process. Four easy steps can produce a large number of plants without requiring fancy equipment.This method of propagation requires that most of our efforts take place in very late autumn or very early spring, two times when we have a short list of gardening chores. The ensuing growing season will find your new plants filling out, and by the fall, they should be ready for a permanent place in your garden design.

Fig. Uncover the Roots when the Plant is Dormant

Root cuttings are most effective if taken when plants are dormant, usually between November and February. This timing maximizes the stored energy in the roots and minimizes the stress on the parent plant. Once a plant breaks bud in the spring, energy begins to move out of the roots and into the plant, reducing the effectiveness of root cuttings. Make sure the parent plant is well hydrated and was not under drought stress when it went dormant before any cuttings. When collecting roots on smaller plants, such as perennials, it easier to simply lift them in their entirety.

For shrubs, some of the roots on one side of the plant, digging in close proximity to the base to ensure I find roots belonging to the right plant, trace them out from the plant until I have a sufficient length to take several cuttings.

Fig. Digging Close to the Base of Shrub.

Fig. A Living Root, not a Dead, Black One.

Fig. Roots as Pencil Thick

Shrub roots can get quite large and woody, but the best ones for cuttings are those approximately as thick as a pencil. These are young, vigourous roots that are more likely to send up new shoots. With perennial roots, thicker is better. Use a sharp pair of pruners to make a straight cut at the end of the root closest to the parent plant. At the far end, make a diagonal cut.This helps maintain the root's original orientation, critical to the production of new roots and shoots, always avoid cutting off more than one-third of the roots because this may eliminate too much of the plant's stored energy, then take the long pieces of root then removed and cut them into sections 3 to 6 inches long, making sure to cut the ends closest to the plant straight and the ends farthest from the plant at an angle.

The optimal length is 3 to 6 inches because it ensures that there is enough energy in the cutting and, in some cases, enough dormant buds to produce roots and shoots. After taking the cuttings, replant the mother plant or cover the exposed roots.

Then water the area thoroughly to remove large air pockets in the soil and settle the roots back into their home. Place the cuttings in soil, and wait To get the most out of cuttings, treat perennials and woody plants differently at this stage.

Fig. Place Perennial Roots Horizontally Because they can send up Shoots from Several Places Along a Cutting.

Fig. Tying the Cuttings in a Bundle makesthem Easier to find when it is time to dig Them up.

Fig. The Entire hole Doesn't need to be Filled with sand. But it is Essential in the Bottom to Reduce the Potential for rot.

Fig. Coloured tape or some other type of Marker Wrapped around the Cuttings and Sticking Aboveground will keep you from Planting on top of the Bundle.

Perennial cuttings grown inside can sprout quicker than their outdoor counterparts. After several shoots emerge,

I remove the protective covers, feed them with a half-strength solution of liquid fertilizer, and, if the plants are sun lovers, supply ample light. Just like seedlings, these new shoots will need to be gradually introduced to the rigors of outdoor life before they can be planted in a permanent home.

Fig. A New Shoot

Fig. Time for Shoots to become Plants. To help the Transition, Provide some TLC in the form of Moist, Fertile, loose Soil.

Fig. Structure of a Plant Water System

A plant's ability to move water through its body and deliver it to cells is an important function for its survival. Beginning in the roots, plants absorb water from the surroundings and use vessels in the stems to transport it upwards. Pressure is the key driving force that allows the water to travel upwards. Even tall trees are able to transport water to their highest leaves.

WATER RELATION

ROOTS

Water is initially absorbed through the plant's roots. Rainwater, run-off, groundwater and other external water sources present in the soil surround the often-long reach of the root system.

Root hairs project outward from the roots, providing additional surface area for optimal water absorption. Once taken into the root hairs, water moves through the root cells that make up the larger root tissue. Passing the series of root cells, water then enters the xylem.

XYLEM

The xylem of a plant is a vascular system that allows water to move up the plant through the stem. The xylem extends from inside the roots, up through the stem, continues into the stalk and ends at the plant leaves. In the leaves, the water continues to travel through increasingly smaller vessels until it reaches its destination.

LEAVES

The final destination for the water traveling through a plant is the plant cell. Here it is used for various plant functions, mainly respiration. Most of the water that is transported through the plant is lost to transpiration, the evaporation of water through leaf pores called stomates. Between 95 and 99 per cent of the water that passes through a plant is lost through transpiration. The remaining 1 per cent to 5 per cent is used in respiration and other cellular operations.

OSMOSIS

Roots absorb water from the soil using a process called osmosis. Osmosis is the transfer of water molecules from a place of high water concentration to a place of low concentration. When the soil around a plant has a higher concentration of water than the plant's root cells, the water molecules pass through the cell walls and enter the plant. The higher the level of attraction a root has for water molecules, the higher its osmotic potential. So when a root cell has a lower water concentration than its surroundings, it is said to have a high osmotic potential.

TRANSPIRATION

The high osmotic potential that attracts water into the root cells begins with the evaporation of water from the plant's leaves. The loss of water resulting from transpiration in the leaves creates a high osmotic potential in the leaf cells. Water is pulled from the xylem cells into the leaf cells. This raises the osmotic potential of the xylem cells. The process continues downward through the plant until water is eventually pulled from the root cells to replace the loss in the lower xylem. The water deficiency created in the roots allows water to come into the plant from the soil. Transpiration, in this way, creates a continual pull of water upward through the plant, replenishing not only the water supply but the nutrient supply as well.

13

Cellular Metabolism and Fermentation

GLYCOLYSIS, THE UNIVERSAL PROCESS

Nine reactions, each catalyzed by a specific enzyme, makeup the process we call glycolysis. ALL organisms have glycolysis occurring in their cytoplasm. At steps 1 and 3 ATP is converted into ADP, inputting energy into the reaction as well as attaching a phosphate to the glucose. At steps 6 and 9 ADP is converted into the higher energy ATP. At step 5 NAD^+ is converted into $NADH^+$ H^+. The process works on glucose, a 6-C, until step 4 splits the 6-C into two 3-C compounds. Glyceraldehyde phosphate (GAP, also known as phospho-glyceraldehyde, PGAL) is the more readily used of the two. Dihydroxyacetone phosphate can be converted into GAP by the enzyme Isomerase. The end of the glycolysis process yields two pyruvic acid (3-C) molecules, and a net gain of 2 ATP and two NADH per glucose.

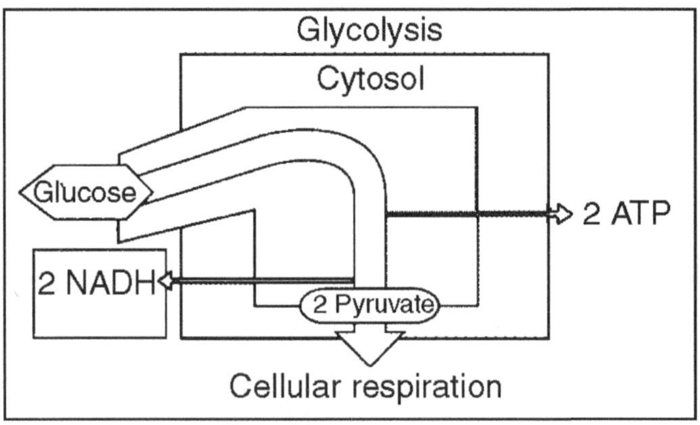

Fig. Glycolysis Process.

ANAEROBIC PATHWAYS

Under anaerobic conditions, the absence of oxygen, pyruvic acid can be routed by the organism into one of three pathways: lactic acid fermentation, alcohol fermentation, or cellular (anaerobic) respiration. Humans cannot ferment alcohol in their own bodies, we lack the genetic information to do so. These biochemical pathways, with their myriad reactions catalyzed by reaction-specific enzymes all under genetic control, are extremely complex. We will only skim the surface at this time and in this course. Alcohol fermentation is the formation of alcohol from sugar. Yeast, when under anaerobic conditions, convert glucose to pyruvic acid via the glycolysis pathways, then go one step farther, converting pyruvic acid into ethanol, a C^{-2} compound.

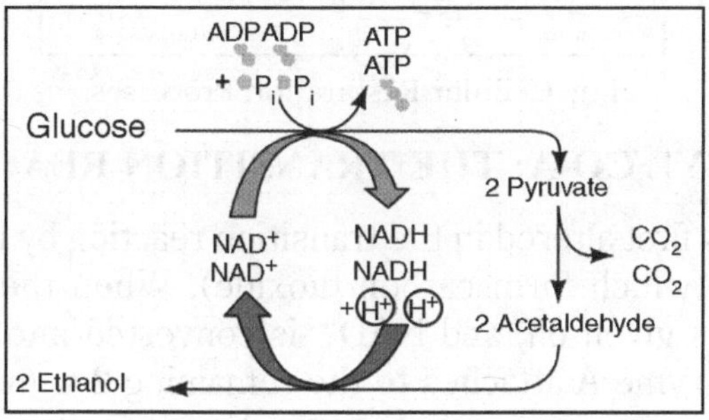

Fig. Fermentation of Ethanol.

Many organisms will also ferment pyruvic acid into, other chemicals, such as lactic acid. Humans ferment lactic acid in muscles where oxygen becomes depleted, resulting in localized anaerobic conditions. This lactic acid causes the muscle stiffness couch-potatoes feel after beginning exercise programmes. The stiffness goes away after a few days since the cessation of strenuous activity allows aerobic conditions to return to the muscle, and the lactic acid can be converted into ATP via the normal aerobic respiration pathways.

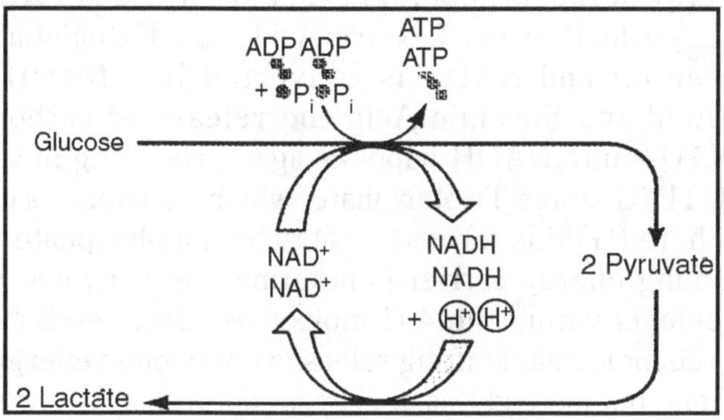

Fig. Fermentation of Lactate (Lactic Acid).

AEROBIC RESPIRATION

When oxygen is present (aerobic conditions), most organisms will undergo two more steps, Kreb's Cycle, and Electron Transport, to produce their ATP. In eukaryotes, these processes occur in the mitochondria, while in prokaryotes they occur in the cytoplasm.

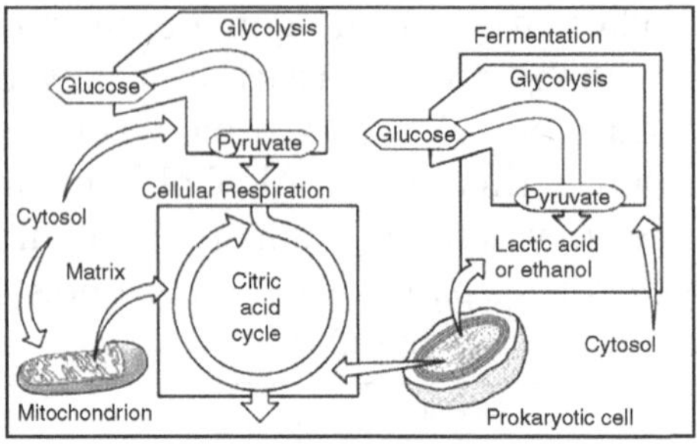

Fig. Cellular Respiration Processes.

ACETYL CO-A: THE TRANSITION REACTION

Pyruvic acid is first altered in the transition reaction by removal of a carbon and two oxygens (which form carbon dioxide). When the carbon dioxide is removed, energy is given off, and NAD^+ is converted into the higher energy form NADH. Coenzyme A attaches to the remaining 2-C (acetyl) unit, forming acetyl Co-A. This process is a prelude to the Kreb's Cycle.

KREB'S CYCLE (AKA CITRIC ACID CYCLE)

The Acetyl Co-A (2-C) is attached to a 4-C chemical (oxaloacetic acid). The Co-A is released and returns to await another pyruvic acid. The 2-C and 4-C make another chemical known as Citric acid, a 6-C. Kreb's Cycle is also known as the Citric Acid Cycle.

The process after Citric Acid is essentially removing carbon dioxide, getting out energy in the form of ATP, GTP, NADH and $FADH_2$, and lastly regenerating the cycle. Between Isocitric Acid and a-Ketoglutaric Acid, carbon dioxide is given off and NAD^+ is converted into NADH. Between a-Ketoglutaric Acid and Succinic Acid the release of carbon dioxide and reduction of NAD+ into NADH happens again, resulting in a 4-C chemical, succinic acid. GTP (Guanine Triphosphate, which transfers its energy to ATP) is also formed here (GTP is formed by attaching a phosphate to GDP).

The remaining energy carrier-generating steps involve the shifting of atomic arrangements within the 4-C molecules. Between Succinic Acid and Fumaric Acid, the molecular shifting releases not enough energy to make ATP or NADH outright, but instead this energy is captured by a new energy carrier, Flavin adenine dinucleotide (FAD). FAD is reduced by the addition of two H's to become $FADH_2$. $FADH_2$ is not as rich an energy carrier as NADH, yielding less ATP than the latter.

The last step, between Malic Acid and Oxaloacetic Acid reforms OA to complete the cycle. Energy is given off and trapped by the reduction of NAD+ to NADH. The carbon dioxide released by cells is generated by the Kreb's Cycle, as are the energy carriers (NADH and $FADH_2$) which play a role in the next step.

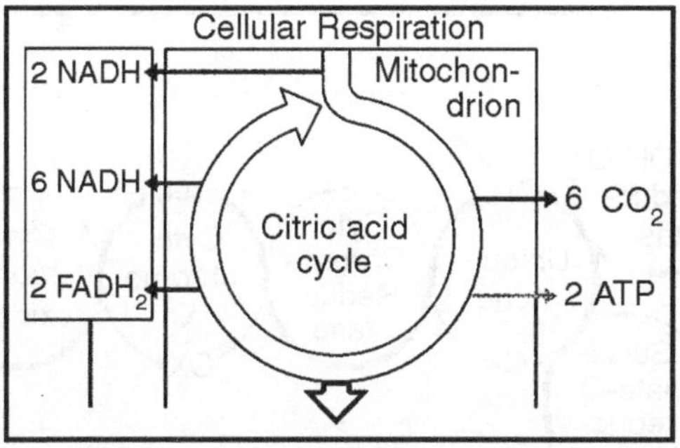

Fig. Summary of the Krebs' (or Citric Acid) Cycle.

ELECTRON TRANSPORT PHOSPHORYLATION

Whereas Kreb's Cycle occurs in the matrix of the mitochondrion, the Electron Transport System (ETS) chemicals are embedded in the membranes known as the cristae. Kreb's cycle completely oxidized the carbons in the pyruvic acids, producing a small amount of ATP, and reducing NAD and FAD into higher energy forms. In the ETS those higher energy forms are cashed in, producing ATP. Cytochromes are molecules that pass the "hot potatoes" (electrons) along the ETS chain. Energy released by the "downhill" passage of electrons is captured as ATP by ADP molecules.

The ADP is reduced by the gain of electrons. ATP formed in this way is made by the process of oxidative phosphorylation. The mechanism for the oxidative phosphorylation process is the gradient of H^+ ions discovered across the inner mitochondrial membrane. This mechanism is known as chemiosmotic coupling.

This involves both chemical and transport processes. Drops in the potential energy of electrons moving down the ETS chain occur at three points. These points turn out to be where ADP + P are converted into ATP. Potential energy is captured by ADP and stored in the pyrophosphate bond. NADH enters the ETS chain at the beginning, yielding 3 ATP per NADH. $FADH_2$ enters at Co-Q, producing only 2 ATP per $FADH_2$.

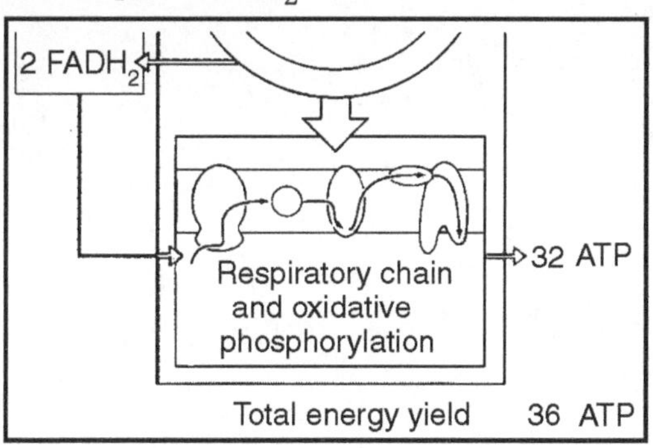

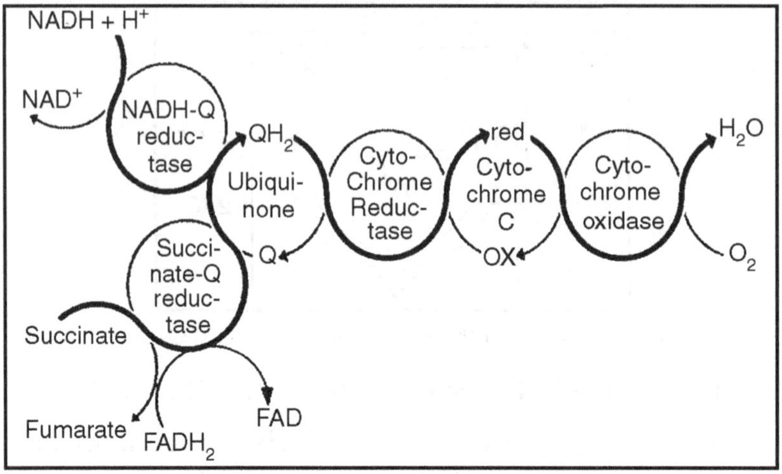

Fig. Electron Transport System.

Bibliography

A B Menan: *Introduction to Plant Anatomy*, Rajat Publication, Delhi, 2008.

A. Hemantaranjan: *Plant Physiology Under Changing Environment*, Scientific Publication, Delhi, 2011.

A. Hemantaranjan: *Plant Physiology, Biochemistry and Plant Molecular Biology in New Millennium*, Scientific Publication, Delhi, 2003.

Aslam Khan: *Plant Anatomy and Physiology*, Kalpaz Publication, Delhi, 2001.

B Politycka and C L Goswami: *Plant Physiology : Research Methods*, Scientific Publication, Delhi, 2007.

B. Mishra, Ramesh Chandra and S.K. Singh: *Management of Soil Quality for Sustainable Agriculture*, Satish Serial Publication, Delhi, 2007.

B. P. Pandey: *Plant Anatomy*, S. Chand Publisher, Delhi, 2001.

Edwin Cyrus Miller: *Plant Physiology : With Reference to the Green Plant (3 Vols-Set)*, Biotech Books, Delhi, 2005.

Edwin Oxlade: *Plant Physiology*, Viva Books, Delhi, 2010.

G. Jahir Hussain: *Introduction to Plant Anatomy*, RBSA Publishers, Jaipur, 2011.

Iqbal Hussain: *Plant Physiology*, Oxford Book Company, Delhi, 2008.

J.K. Maheshwari: *Recent Researches in Plant Anatomy and Morphology*, Scientific Publication, Delhi, 2001.

J.K.A. Bleasdale: *Plant Physiology in Relation to Horticulture*, Scientific Publication, Delhi, 2014.

Kabita Dutta: *Plant Physiology*, International Scientific Publication, Delhi, 2011.

Mohammed Shamsuddin: *Silkworm Physiology : A Concise Textbook*, Daya Publication, Delhi, 2009.

Mohr: *Plant Physiology*, Springer, New York, 2001.

Neeru Mathur: *Plant Physiology*, RBSA Publishers, Jaipur, 2010.

P C Trivedi: *Plant Physiology : Current Trends*, Pointer Publication, Delhi, 2007.

P C Trivedi; Anukriti Atreya and Kirti Pathak: *Plant Physiology, Biochemistry and Biotechnology*, Indus Valley, Delhi, 2006.

P.C. Trivedi: *Plant Physiology in Agriculture and Forestry*, Aavishkar Publication, Delhi, 2009.

Rajaram Choyal: *Plant Anatomy and Physiology*, Sonali Publications, Delhi, 2012.

Ramesh Chandra and Satish Kumar Singh: *Fundamentals and Management of Soil Quality*, Westville Publication, Delhi, 2009.

Renuka Desai: *Plant Anatomy : Principles and Practices*, Adhyayan Publication, Delhi, 2008.

Renuka Desai: *Plant Physiology*, Adhyayan Publication, Delhi, 2008.

S.K. Singh and Seema Srivastava: *Plant Physiology and Biochemistry*, Campus Books, Delhi, 2009.

S.S. Purohit and A. Kumar: *Plant Physiology Fundamentals and Applications*, Agrobios Publication, Delhi, 2011.

Susheela M. Das and Abha Bhardwaj: *Plant Physiology*, Wisdom Press, Delhi, 2012.

Susheela M. Das: *Latest Portfolio of Theory and Practice in Plant Anatomy*, Dominant Publication, Delhi, 2003.

Index